BRIDGE ENGINEERING

BRIDGE ENGINEERING

Classifications, Design Loading, and Analysis Methods

WEIWEI LIN
TERUHIKO YODA

Butterworth-Heinemann
An imprint of Elsevier

Butterworth-Heinemann is an imprint of Elsevier
The Boulevard, Langford Lane, Kidlington, Oxford OX5 1GB, United Kingdom
50 Hampshire Street, 5th Floor, Cambridge, MA 02139, United States

Notices
Knowledge and best practice in this field are constantly changing. As new research and experience broaden our understanding, changes in research methods, professional practices, or medical treatment may become necessary.

Practitioners and researchers must always rely on their own experience and knowledge in evaluating and using any information, methods, compounds, or experiments described herein. In using such information or methods they should be mindful of their own safety and the safety of others, including parties for whom they have a professional responsibility.

To the fullest extent of the law, neither the Publisher nor the authors, contributors, or editors, assume any liability for any injury and/or damage to persons or property as a matter of products liability, negligence or otherwise, or from any use or operation of any methods, products, instructions, or ideas contained in the material herein.

Library of Congress Cataloging-in-Publication Data
A catalog record for this book is available from the Library of Congress

British Library Cataloguing-in-Publication Data
A catalogue record for this book is available from the British Library

ISBN: 978-0-12-804432-2

Working together
to grow libraries in
developing countries

www.elsevier.com • www.bookaid.org

Publisher: Matthew Deans
Acquisition Editor: Ken McCombs
Editorial Project Manager: Peter Jardim
Production Project Manager: Anusha Sambamoorthy
Cover Designer: Mark Rogers

Typeset by SPi Global, India

CONTENTS

ABOUT THE AUTHORS

Weiwei Lin is a member of the Department of Civil and Environmental Engineering and International Center for Science and Engineering Programs (ICSEP), Waseda University, holding associate professorship in the Bridge Engineering Laboratory. He has authored or coauthored over 100 academic papers, proceedings, and technical articles dealing with the problems of structural mechanics and bridge engineering, especially for the steel structures and steel-concrete composite structures. He is a member of several engineering committees, like ASCE, JSCE, IABSE, IABMAS, IALCCE, etc. He is also the recipient of IABMAS YOUNG PRIZE of 2014.

Teruhiko Yoda is on the faculty of Waseda University, where he holds the chair professorship in the Department of Civil and Environmental Engineering. He has authored or coauthored 7 technical books and over 400 articles dealing with the problems of structural mechanics and bridge engineering. He is a member of the ASCE, JSCE, and IABSE and former chairman of International Committee of JSCE, and the former president of Kanto Branch of JSCE. Besides, he is chairman of the Drafting Committee for Standard Specifications for Steel and Composite Structures (First Edition 2007). He is the recipient of many Japanese awards, including the prestigious Tanaka Award.

CHAPTER ONE

Introduction of Bridge Engineering

1.1 INTRODUCTION

A bridge is a construction made for carrying the road traffic or other moving loads in order to pass through an obstacle or other constructions. The required passage may be for pedestrians, a road, a railway, a canal, a pipeline, etc. Obstacle can be rivers, valleys, sea channels, and other constructions, such as bridges themselves, buildings, railways, or roads. The covered bridge at Cambridge in Fig. 1.1 and a flyover bridge at Osaka in Fig. 1.2 are also typical bridges according to above definition. Bridges are important structures in modern highway and railway transportation systems, and generally serving as "lifelines" in the social infrastructure systems.

Bridge engineering is a field of engineering (particularly a significant branch of structural engineering) dealing with the surveying, plan, design, analysis, construction, management, and maintenance of bridges that support or resist loads. This variety of disciplines requires knowledge of the science and engineering of natural and man-made materials, composites, metallurgy, structural mechanics, statics, dynamics, statistics, probability theory, hydraulics, and soil science, among other topics (Khan, 2010). Similar to other structural engineers (Abrar and Masood, 2014), bridge engineers must ensure that their designs satisfy given design standard, being responsible to structural safety (i.e., bridge must not deform severely or even collapse under design static or dynamic loads) and serviceability (i.e., bridge sway that may cause discomfort to the bridge users should be avoided). Bridge engineering theory is based upon modern mechanics (rational knowledge) and empirical knowledge of different construction materials and geometric structures. Bridge engineers need to make innovative and high efficient use of financial resources, construction materials, calculation, and construction technologies to achieve these objectives.

Bridge Engineering
http://dx.doi.org/10.1016/B978-0-12-804432-2.00001-3

Fig. 1.1 The Bridge of Sighs, Cambridge, the United Kingdom. *(Photo by Lin.)*

Fig. 1.2 A flyover in Osaka, Japan. *(Photo by Lin.)*

1.2 BRIDGE COMPONENTS

1.2.1 Superstructure, Bearings, and Substructure

Structural components of bridges are based on parametric definitions involving deck types and various bridge properties. Bridge structures are composed of superstructure, bearing, superstructure, and accessories.

(A) Superstructure

In general, the superstructure represents the portion of a bridge above the bearings, as shown in Fig. 1.3. Superstructure is the part of a bridge supported by the bearings, including deck, girder, truss, etc. The deck directly carries traffic, while other portions of the superstructure bear the loads passing over it and transmit them to the substructures. In case, the deck was divided as a separate bridge component, and the structural members between the deck and the bearings are called as bridge superstructure.

The superstructure may only include a few components, such as reinforced concrete slab in a slab bridge, or it may include several components, such as the floor beams, stringers, trusses, and bracings in a

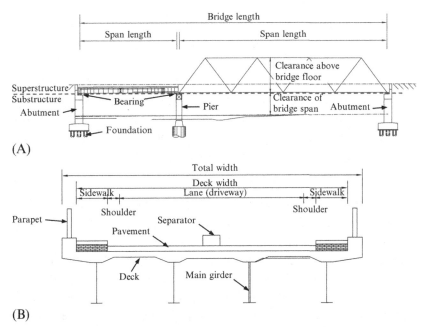

Fig. 1.3 General terminology of bridges. (A) Longitudinal direction. (B) Cross section.

truss bridge. In suspension and cable–stayed bridges, components such as suspension cables, hangers, stays, towers, bridge deck, and the supporting structure comprise the superstructure (Taly, 1997).

(B) Bearings

A bridge bearing is a component of a bridge transmitting the loads received from the deck on to the substructure and to allow controlled movement due to temperature variation or seismic activity and thereby reduce the stresses involved. A bearing is the boundary between the superstructure and the substructure.

(C) Substructure

Substructure is the portion of the bridge below the bearing, used for supporting the bridge superstructure and transmits all those loads to ground. In this sense, bridge substructures include abutments, piers, wing walls, or retaining walls, and foundation structures like columns and piles, drilled shafts that made of wood, masonry, stone, concrete, and steel.

Both abutments and piers are vertical structures used for supporting the loads from the bridges bearings or directly from the superstructures and for transmitting the load to the foundation. However, the abutments refer to the supports located at beginning or end of bridge, while the piers are the intermediate supports. Therefore, a bridge with a single span has only abutments at both ends, while multispan bridges also need intermediate piers to support the bridge superstructures, as can be seen in Fig. 1.3.

(D) Accessory structures

Bridge accessories are structure members subordinate to the main bridge structure, such as parapets, service ducts, and track slabs. Deadweight of accessory structures shall be considered in the design, but their load carrying capacities are generally ignored.

1.2.2 Bridge Length, Span Length, and Bridge Width

The distance between centers of two bearings at supports is defined as the span length or clear span. The distance between the end of wing walls at either abutments or the deck lane length for bridges without using abutments is defined as total bridge length. Obviously, the bridge length is different from the span length. For example, the world's largest bridge (means the span length) is the Akashi Kaikyō Bridge in Japan (with the central span of 1991 m), while the longest bridge (means the total length) is the

Danyang-Kunshan Grand Bridge in China, which is a 164.8-km long via-duct on the Beijing-Shanghai High-Speed Railway.

Deck width is the sum of the carriageway width, sidewalk width, shoulder width, and the individual elements required to make up the desired bridge cross section. The total bridge width not only includes the deck width but also the width of the bridge accessories such as parapets. The lane width is determined according to the bridge design codes, generally with the minimum width of 2.75 m and the maximum width of 3.5 m.

1.2.3 Bridge Clearance

There are two types of bridge clearance, including clearance of bridge span and clearance above bridge floor. Clearance of bridge span is generally measured from the water surface (or ground, if there is no water) to the under-surface of the bridge. The measurement from the mean highest high water (MHHW) is the most conservative clearance, thus in most cases the real clearance is larger than this value due to the lower water surface than the highest point at MHHW. Enough clearance should be considered in the bridge design to ensure the traffic safety under the bridge. Clearance above bridge floor is the space limit for carriageway and sidewalk, which is generally specified in the bridge design specification to ensure the traffic safety (enough height or space) above the bridge.

1.3 BRIDGE CLASSIFICATION

Depending on the objective of classification, the bridges can be classified in several ways. The necessity of classifying bridges in various ways has grown as bridges have evolved from simple beam bridges to modern cable-stayed bridges or suspension bridges. Bridges are always classified in terms of the bridge's superstructure, and superstructure can be classified according to the following characteristics:

Materials of construction
Span length
Position (for movable bridges)
Span types
Deck location
Usage
Geometric shape
Structural form

1.3.1 Bridge Classification by Materials of Construction

Bridges can be identified by the materials from which their superstructures are built, namely, steel, concrete, timber, stone, aluminum, and advanced composite materials. This is not suggested that only one kind of material is used exclusively to build these bridges. Frequently, a combination of materials is used in bridge building. For example, a bridge may have a reinforced concrete deck and steel main girders, which is typically used in highway bridge superstructures. New materials such as advanced composite materials have also been widely used in bridge construction.

1.3.2 Bridge Classification by Span Length

In practice, it is general to classify bridges as short span, medium span, and long span, according to their span lengths. The concept of "super-long span bridges," defining a bridge with a span much longer than any existing bridges, was also proposed in recent years (Tang, 2016). However, up to now there are no standard criteria to define the range of spans for these different classifications. A criterion proposed by Taly (1997) is to classify bridges by span length as follows:

Culverts	$L \leq 20$ ft ($\sim$6 m)
Short-span bridges	20 ft $< L \leq 125$ ft (approximately from 6 to 38 m)
Medium-span bridges	125 ft $< L \leq 400$ ft (approximately from 38 to 125 m)
Long-span bridges	$L > 400$ ft (125 m $\sim$)

As already discussed above, this is an often used but not a standard criterion. Taking the long span as an example, it was also proposed that a span length less than or equal to 180 (Lutomirska and Nowak, 2013) or 200 m (Catbas et al., 1999). The current bridge design specification for highway bridges in Japan is applicable for a bridge with a span length <200 m or less. At this point, it seems more reasonable to define a long-span bridge in Japan as a span length up to 200 m, but not 125 m. This is reasonable because the span capacity of a bridge depends on many factors, such as their structural form, construction materials, design methods, and construction techniques. For instance, the span of a girder bridge cannot be compared with the span of a cable-stayed bridge in length, and also a bridge classified as long span nowadays may be changed to medium span in the future.

This classification of bridges according to span length is made more for the sake of description, which was useful for bridge type selection. In general, certain types of bridges are suitable only for a certain range of span lengths. For example, a suspension bridge or a cable-stayed bridge is generally used for long spans, thus it should not be considered as an alternative for a short-span bridge. Similarly, a bridge type suitable for short-span bridges (such as a type of beam bridge) should not be used for bridges with long spans.

1.3.3 Bridge Classification by Position-Moveable Bridges

A moveable bridge is a bridge that moves to allow passage usually for boats or barges (Schneider, 1907). An advantage of making bridges moveable is the lower construction cost due to the absence of high piers and long approaches. Three types often used moveable bridges are bascule bridges, swing bridges, and lift bridges.

1.3.3.1 Bascule Bridges

A bascule bridge is a kind of widely used moveable bridge whose main girders can be lifted together with deck about the hinge located at the end of the span. Depending on the bridge width, the bascule bridge can be designed as either single or double leafed. Tower Bridge (built 1886–94) crosses the River Thames in London is a combined suspension bridge and bascule, as shown in Fig. 1.4.

1.3.3.2 Swing Bridges

In swing bridges, the girders together with the deck can be swung about the vertical support ring at the pier in the middle (or abutment at the end), to

Fig. 1.4 The Tower Bridge in London. *(Photo by Yoda.)*

Fig. 1.5 Two swing bridges in Liverpool. *(Photos by Lin.)*

allow the traffic to cross. Small swing bridges may be pivoted only at one end, opening like a gate, but require substantial base structure to support the pivot. Two swing bridges in Liverpool are shown in Fig. 1.5.

1.3.3.3 Lift Bridges

In lift bridges, gantries are provided at the piers at either end of the span. Both girder and the floor system are lifted up by a hydraulic arrangement to the extent required for free passage of the ship (Ponnuswamy, 2008). The Stillwater Lift Bridge shown in Fig. 1.6 is a typical bridge of this type.

In addition to those moveable bridges mentioned above, drawbridges, folding bridges, retractable bridges, curling bridges, tilt bridges, and Jet bridges are also usually used. However, in comparison with other bridges, the moveable bridges are generally characterized as higher inspection and maintenance costs, difficult to widen in the future, and poor seismic performance.

1.3.4 Bridge Classification by Interspan Relation

According to the interspan relations, generally the bridge structures can be classified as simply supported, continuous, or cantilever bridges, as shown in Fig. 1.7.

Fig. 1.6 A lift bridge in Minnesota (the Stillwater Lift Bridge). *(Photo by Yoda.)*

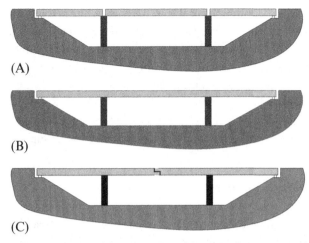

Fig. 1.7 Simply supported, continuous, and cantilever bridges. (A) Simply supported span. (B) Continuous span. (C) Cantilever span.

1.3.4.1 Simply Supported Bridges

For this type of bridge, the load carrying member is simply supported at both ends. They are statically determinate structures and suitable to be constructed at bridge foundations that uneven settlements are likely to happen. In general, the bridge is divided into several individual spans with relatively short-span length. Due to the maximum bending moment at the mid span and maximum shear force at girder ends, simply supported bridges are generally designed with constant girder height to simplify the design and construction.

1.3.4.2 Continuous Bridges

Continuous bridges are statically indeterminate structures, whose spans are continuous over three or more supports. In comparison with simply supported girder bridges, the continuous bridges have been used extensively in bridge structures due to the benefits of higher span-to-depth ratio, higher stiffness ratios, reduced deflections, less expansion joints, and less vibration. In continuous bridges, the positive bending moment is much smaller than that in simply supported span due to the absence of the negative bending at the intermediate piers; thus they generally need smaller sections and have considerable saving compared to simply supported bridge construction. Due to the relatively large negative bending moment and shear forces at intermediate supporting sections, larger girder depth than that in span center section is generally used.

In addition, the continuous bridge requires only one bearing at each pier as the bearings which can be placed at the center of piers in comparison with two bearings for a simply supported bridge, and the reactions at piers are transmitted centrally. However, the continuous bridges also have some disadvantages, such as the design is more complicated because they are statically indeterminate. In the negative bending moment zone, concrete deck is easy to crack while the bottom steel girder is vulnerable to buckling. Also, large internal forces may occur due to temperature variation or uneven settlement of supports.

1.3.4.3 Cantilever Bridges

The cantilever bridge is a bridge whose main structures are cantilevers, which are used to build girder bridges and truss bridges. A cantilever bridge has advantages in both simply supported and continuous bridges, like they are suitable for foundation with uneven settlement; they can be built without false-works but has larger span capacity. For cantilever bridges with balanced construction, hinges are usually provided at contra flexure points of a continuous span, and an intermediate simply supported span can be suspended between two hinges. Cantilever bridges were not only built as girder bridges but also widely used in truss bridges. The Quebec Bridge in Canada and the Forth Bridge in United Kingdom (Fig. 1.8) are the top two largest cantilever truss bridges in the world.

Fig. 1.8 The Forth Bridge in Scotland. *(Photo by An.)*

1.3.5 Bridge Classification by Deck Location

According to the relative location between the bridge deck and the main (load carrying) structure, the bridge superstructures are classified as deck bridges, through bridges, and half-through bridges. The bridge is defined as a deck bridge when the deck is placed on the top of the main structure. If the deck is located on the bottom of the main structure, it is a through bridge. While, if the deck is located on the middle of the main structure, it is a half-through bridge. Considering the traffic, the bridges should be built as a deck bridge if possible. As a special case, the bridge will be classified as a double-deck bridge if two layers of deck are used.

1.3.6 Bridge Classification by Geometric Shape

According to the geometric shape, the bridge superstructures can be classified as straight (or right) bridges, skew bridges, and curved bridges, as shown in Fig. 1.9.

1.3.6.1 Straight Bridges

If the bridge axis follows a straight line, then it is a straight bridge, as shown in Fig. 1.9A. The bridges should be constructed in straight to avoid the extra forces such as torsions and to simplify the bridge design, analysis, and construction.

1.3.6.2 Skewed Bridges

Skewed bridges (Fig. 1.9B) are often used in highway design when the geometry cannot accommodate straight bridges. Skewed bridges are generally not preferred and sparingly chosen due to the difficulties in the design. However, it is sometimes not possible to arrange that a bridge spans square to the feature that it crosses, particularly where it is necessary to keep a straight

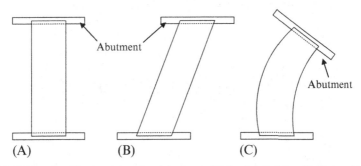

Fig. 1.9 Bridge classification by geometric shape. (A) Straight bridge. (B) Skewed bridge. (C) Curved bridge.

alignment of a roadway above or below the bridge. On this occasion, a skew bridge is required.

In AASHTO LRFD Bridge Design Specifications (2004), it is suggested that skew angles under 15 degrees can be ignored. While for skew angles larger than 30 degrees, the effects of skew angles are usually considered significant and need to be considered in analysis. The torsional effects due to the skew support arrangements must be taken into account in design. Skewed bridges have a tendency to rotate under seismic loading, thus bearings should be designed and detailed to accommodate this effect.

1.3.6.3 Curved Bridges

In comparison with a straight bridge, a curved bridge is more difficult in both design and construction. Most highway and railway bridges follow a straight alignment, while some bridges need to be designed as partly or wholly curved in plan for different purposes. For road bridges, like interconnected urban vehicular overpasses, curvature is usually required for the convenience in spatial arrangement. For pedestrian bridges, curvature may be employed either for providing users a unique spatial experience, to bring them into unattainable locations, or for esthetic purposes. A good example of such bridges is the Langkawi Sky Bridge built on the Machinchang Mountain top in Malaysia, as shown in Fig. 1.10.

Like the skew bridges, the bearing arrangements in curved bridges also need to be carefully designed.

Fig. 1.10 Langkawi Sky Bridge.

1.3.7 Bridge Classification by Usage

A bridge can be categorized by what it is designed to carry, such as road traffic, rail traffic, pedestrian, a pipeline or waterway for barge traffic, or water transport. According to the utility (or function), bridges can be classified into highway bridges, railway bridges, pedestrian bridges, aqueduct bridges, pipeline bridges, airport runway bridges, combined bridges, etc. Highway bridges are designed for vehicle load, pedestrian load, and other loads, while a railway bridge (e.g., a steel trestle railway bridge shown in Fig. 1.11) is built mainly for carrying railroad traffic, either cargo or passenger. A road-rail bridge designed as double deck carries both road and rail traffic. In addition to highway and railway bridges, there are some other bridges designed to carry nonvehicular traffic and loads. These bridges include pedestrian bridge, airport runway bridge, aqueduct bridge, pipeline bridge, and conveyor bridges.

A pedestrian bridge (or referred to as a footbridge) is designed for pedestrians, cyclists, or animal traffic, rather than vehicular traffic. In many cases, footbridges are both beautiful works of art and functional as a bridge. Millennium Footbridge in London and Lagan Weir Footbridge in Belfast are two beautiful footbridges in the United Kingdom, as shown in Figs. 1.12 and 1.13, respectively. An airport runway bridge is built as runways for airplanes, and its width mainly depends on the wingspan of the aircraft, which varies widely. The design of the airport runway bridge depends on the weight, the landing gear pattern, and the wingspan (Taly, 1997).

An aqueduct bridge is a bridge constructed for carrying water, like a viaduct that connects points of same height. The famous Aqueduct Bridge in

Fig. 1.11 The former Amarube Bridge (a steel trestle railway bridge). *(Photo by Yoda.)*

Fig. 1.12 London Millennium Footbridge. *(Photo by Lin.)*

Fig. 1.13 Lagan Weir Footbridge. *(Photo by Lin.)*

Fig. 1.14 The Aqueduct Bridge. *(Photo by An.)*

Spain is a representative bridge of this type, as shown in Fig. 1.14. Pipeline bridges are designed for carrying the fluids such as water, oil, and gas when it is not possible to run the pipeline on a conventional bridge or under the river, like those shown in Fig. 1.15. A walkway may be equipped in a pipeline bridge for maintenance purposes. But, in most cases, this is not open for public access for security reasons. In addition, a conveyor bridge is designed as an automatic unit for the removal of overburden and for dumping it onto the inner spoil banks of open cut mines.

A combined bridge is designed for two or more functions. In addition, temporary bridges that are used in natural disasters (also named as emergency bridges) and in the war (military bridges) that can be easily assembled and then taken apart in the war are also used in practice. On the contrary, the bridges used for long periods are defined as permanent bridges.

1.3.8 Bridge Classification by Structural Form

Although bridges can be classified by different methods, the bridge classification according to its structural form is still the common way. This is necessary because the structural form is the most important factor that affects the whole service life of the bridge, including design, construction, repair, and maintenance. Bridges with different structural forms have their load transfer path and suitable range of application. In general, bridges can be classified into beam bridges, rigid-frame bridges, truss bridges, arch bridges, cable-stayed bridges, and suspension bridges.

Fig. 1.15 Pipeline bridges. *(Photos by Lin.)*

1.3.8.1 Beam Bridges

Beam bridges (also referred to as Girder Bridges) are the most common, inexpensive, and simplest structural forms supported between abutments or piers. In its most basic form, a beam bridge is just supported at each end by piers (or abutments), such as a log across a creek. The weight of the beam and other external load need to be resisted by the beam itself, and the internal forces include the bending moment and shear force. When subjected a positive bending moment, the top fibers of a beam are in compression (pushed together) while the bottom fibers are in tension (stretched). This is more complex than a cable only in tension or an arch mainly in compression. Therefore, only materials that can work well for both tension and compression can be used to build a beam bridge. Obviously, both plain concrete and stone are not good materials for a beam because they are strong in compression, but weak in tension. Though ancient beam bridges were mainly made of wood, modern beam bridges can also be made iron, steel,

Fig. 1.16 Lagan bridge (concrete continuous girder bridge), Belfast.

Fig. 1.17 Queen Elizabeth II Bridge (steel continuous girder bridge), Belfast.

or concrete with the aid of prestressing. Two continuous girder bridges that made of steel and concrete are shown in Figs. 1.16 and 1.17.

Sometimes, the beam bridges are also classified into slab bridges, beam bridges, and girder bridges. As noted by Smith et al. (1989), the slab bridges refer to spans without support below the deck, Beam Bridges represents bridges with only longitudinal support below the deck and Girder Bridges refer to bridges with both longitudinal and transverse structural members under the deck. In this book, however, all these three categories will be classified as the same type because of their similar load transfer mechanisms.

1.3.8.2 Rigid-Frame Bridges

A Rigid-Frame Bridge (also known as Rahmen Bridge) consists of superstructure supported on vertical or slanted monolithic legs (columns), in which the superstructure and substructure are rigidly connected to act as a unit and are economical for moderate medium–span lengths. The use of rigid-frame bridges began in Germany in the early 20th century.

(A)

(B)

Fig. 1.18 The second Shibanpo Bridge in Chongqing, China. (A) Layout of the bridge. (B) Main span after construction. *(Photos by Yan.)*

The rigid-frame bridges are superstructure–substructure integral structures with the superstructure can be considered as a girder. Bridges of superstructure–substructure integral structure include braced rigid-frame bridges, V-leg rigid-frame bridges, and viaducts in urban areas. The connections between superstructure and substructure are rigid connections which transfer bending moment, axial forces, and shear forces. A bridge design consisting of a rigid frame can provide significant structural benefits but can also be difficult to design and construct. Moments at the center of the deck of a rigid-frame bridge are smaller than the corresponding moments in a simply supported deck. Therefore, a much shallower cross section at mid-span can be used. Additional benefits are that less space is required for the approaches and structural details for where the deck bears on the abutments are not necessary (Portland Cement Association, 1936). However, as a statically indeterminate structure, the design and analysis is more complicated than that of simply supported or continuous bridges. Spanning $(86.5 + 4 \times 138 + 330 + 132.5)$ m across the Yangzi River (Fig. 1.18), the continuous prepressed rigid-frame Chongqing Shibanpo double-line Bridge, has a world record main span of 330 m in its category (Qin et al., 2013). The Toosu Bridge in Tokyo is also a typical rigid-frame bridge, as shown in Fig. 1.19.

1.3.8.3 Truss Bridges

Truss is a structure of connected elements forming triangular units, and a bridge whose load-bearing superstructure is composed of a truss is a truss

Fig. 1.19 The Toyosu Bridge in Tokyo, Japan. *(Photo by Zheng.)*

bridge. Truss bridges are one of the oldest types of modern bridges. In order to simplify the calculation, trusses are generally assumed as pinned connection between adjacent truss members. Therefore, the truss members like chords, verticals, and diagonals act only in either tension or compression. For modern truss bridges, gusset plate connections are generally used, then bending moments and shear forces of members should be considered for evaluating the real performance of the truss bridges, which is achieved by the aid of finite element software. For the design point of view, however, the pinned connection assumption is considered for security concerns and also for simplifying the structural design and analyses. In addition, as the axial forces (but not bending moments and shear forces) are generally governs the stress conditions of the members, such assumption generally will not cause large errors between the real bridges and the design models.

According to this assumption, the truss members can be in tension, compression, or sometimes both in response to dynamic loads. Typical axial forces in truss members in Pratt truss and Warren truss under deadweight are shown in Fig. 1.20. Owing to its simple design method and efficient use of materials, a truss bridge is economical to design and construct.

Short-span truss bridges are built as simply supported, while the large span truss bridges are generally built as continuous truss bridges or cantilever truss bridges. The list of longest truss bridges in the world is shown in Table 1.1, indicating that most of the large span truss bridges were built as cantilever. The structural features of truss bridges will be discussed in Chapter 8.

The maximum single span of the continuous truss bridge is 440 m in Tokyo Gate Bridge in Japan, as shown in Fig. 1.21. This bridge spans a major sea lane into Tokyo Bay, but its height had to be restricted because it is located near the Haneda Airport. For this reason, other designs alternatives such as suspension bridge and cable-stayed bridge, etc. which needs relative

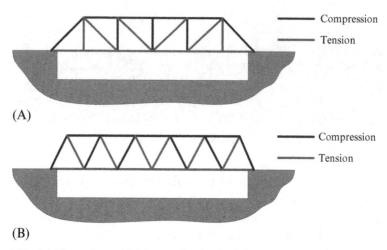

Fig. 1.20 Axial forces in truss bridges under deadweight. (A) Pratt truss. (B) Warren truss.

high towers were repudiated. Although it can be designed as cantilever like other truss bridges shown in Table 1.1, the structural form of continuous truss was selected for the sake of good seismic performance in the seismically active area.

1.3.8.4 Arch Bridges

An arch bridge is a bridge shaped as an upward convex curved arch to sustain the vertical loads. A simple arch bridge works by transferring its weight and other loads partially into a horizontal thrust restrained by the strong abutments at either side. The arch rib needs to carry bending moment, shear force, and axial force in real service conditions. A viaduct (a long bridge) may be made from a series of arches although other more economical structures are typically used today. The current world's largest arch bridge is the Chaotianmen Bridge over the Yangtze River in Chongqing (China) with a span length of 552 m, as shown in Fig. 1.22.

For statically indeterminate arch bridges, the internal forces will occur due to the temperature variation and settlement of supports. For this reason, if the arch bridges are constructed in soft soil foundations, the bridge deck is generally designed to sustain the horizontal forces. Such arch bridges can be found in Fig. 1.23 (Hayashikawa, 2000). More details about arch bridges will be discussed in Chapter 9 (Table 1.2).

Table 1.1 List of Longest Truss Bridges

Rank	Name	Main Span (m)	Year Opened	Location	Country	Type
1	Québec Bridge	549	1917	Quebec	Canada	Cantilever
2	Forth Bridge	521	1890	Scotland	United Kingdom	Cantilever
3	Minato Bridge	510	1973	Osaka	Japan	Cantilever
4	Commodore Barry Bridge	501	1974	New Jersey	United States	Cantilever
5	Crescent City Connection	480	1958 (eastbound), 1988 (westbound)	New Orleans, Louisiana	United States	Cantilever
6	Howrah Bridge	457	1943	West Bengal	India	Cantilever
7	Veterans Memorial Bridge	445	1995	Louisiana	United States	Cantilever
8	Tokyo Gate Bridge	440	2012	Tokyo	Japan	Continuous
9	San Francisco–Oakland Bay Bridge	427	1936	California	United States	Cantilever
10	Ikitsuki Bridge	400	1991	Nagasaki Prefecture	Japan	Continuous

Fig. 1.21 The Tokyo Gate Bridge. *(Photo by Lin.)*

Fig. 1.22 The Chaotianmen Bridge, Chongqing, China. *(Photo by Yan.)*

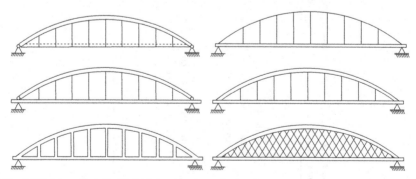

Fig. 1.23 Internal statically indeterminate arch structures.

Table 1.2 List of Longest Arch Bridges

Rank	Name	Main Span (m)	Material	Year Opened	Location	Country
1	Chaotianmen Bridge	552	Steel	2009	Chongqing	China
2	Lupu Bridge	550	Steel	2003	Shanghai	China
3	Bosideng Bridge	530	CFST	2012	Hejiang County (Sichuan)	China
4	New River Gorge Bridge	518	Steel	1977	Fayetteville (West Virginia)	United States
5	Bayonne Bridge	510	Steel	1931	Kill Van Kull (New York)	United States
6	Sydney Harbour Bridge	503	Steel	1932	Sydney	Australia
7	Wushan Bridge	460	CFST	2005	Wushan (Chongqing)	China
8	Xijiang Railway Bridge	450	Steel	2014	Zhaoqing (Guangdong)	China
9	Mingzhou Bridge	450	Steel	2011	Ningbo (Zhejiang)	China
10	Zhijinghe River Bridge	430	CFST	2009	Dazhipingzhen (Hubei)	China

Note: CFST denotes the concrete filled steel tubular structure.

1.3.8.5 Cable-Stayed Bridges

A cable-stayed bridge is a structure with several points in each span between the towers supported upward in a slanting direction with inclined cables and consists of main tower(s), cable-stays, and main girders, as shown in Fig. 1.24. In comparison with the continuous girder bridges, the internal forces due to both dead load and live load are much smaller in cable-stayed bridges. For mechanical point of view, a cable-stayed bridge is a statically indeterminate continuous girder with spring constraints. The cable-stayed

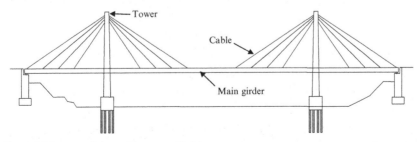

Fig. 1.24 Image of the cable-stayed bridge.

Table 1.3 List of Longest Cabled-Stayed Bridges

Rank	Name	Main Span (m)	Year Opened	Location	Country
1	Russky Bridge	1104	2012	Vladivostok	Russia
2	Sutong Bridge	1088	2008	Suzhou-Nantong	China
3	Stonecutters Bridge Yi-Stonecutters Island	1018 Hong Kong	2009	Tsing	
4	E'dong Bridge	926	2010	Huangshi	China
5	Tatara Bridge	890	1999	Ikuchi Island–Ōmishima Island	Japan
6	Pont de Normandie	856	1995	Le Havre–Honfleur	France
7	Jiujiang Fuyin Expressway Bridge	818	2013	Jiujiang–Huangmei	China
8	Jingyue Bridge	816	2010	Jingzhou–Yueyang	China
9	Incheon Bridge	800	2009	Incheon–Yeongjongdo	South Korea
10	Xiamen Zhangzhou Bridge	780	2013	Zhangzhou–Xiamen	China

bridges are also highly efficient in use of materials due to their structural members mainly works in either tension or compression (axial forces). The details of the cable–stayed bridges will be discussed in Chapter 10.

Cable-stayed bridges have the second-longest spanning capacity (after suspension bridges), and they are practically suitable for spans up to around 1000 m. The top 10 largest cable-stayed bridges are listed in Table 1.3. The Russky Bridge in Russia has the largest span of 1104 m. It is longer by 16 m than the second place Sutong Bridge (largest span is 1088 m) over the Yangtze River in China. The Tatara Bridge with the center span of 890 m, as shown in Fig. 1.25, is currently the largest cable-stayed bridge in Japan and the fifth longest main span of any cable-stayed bridge in the world.

1.3.8.6 Suspension Bridges

A typical suspension bridge is a continuous girder suspended by suspension cables, which pass through the main towers with the aid of a special structure known as a saddle, and end on big anchorages that hold them. Fig. 1.26 shows the essential structural members and elements of typical, including tower, hanger, main girder, and the anchorage. The main forces in a

Fig. 1.25 The Tatara Bridge. *(From https://commons.wikimedia.org/wiki/File:TataraOhashi. jpg.)*

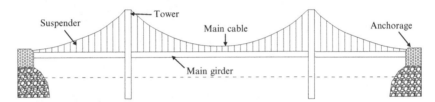

Fig. 1.26 Image of the suspension bridge.

suspension bridge are tension in the cables and compression in the towers. The deck, which is usually a truss or a box girder, is connected to the suspension cables by vertical suspender cables or rods, called hangers, which are also in tension. The weight is transferred by the cables to the towers, which in turn transfer the weight to the anchorages on both ends of the bridge, then finally to the ground.

The curve shape of the suspension cables is similar to that of arch. However, the suspension cable can only sustain the tensile forces, which is different from the compressive forces in the arch. Also because of this, the cable will never "buckle" and highly efficient use of high strength steel materials becomes possible. The use of suspension bridges makes longer main spans achievable than with any other types of bridges, and they are practical for spans up to around 2 km or even larger. The top 10 largest suspension bridges in the world are listed in Table 1.4. The Akashi Kaikyō

Table 1.4 List of Longest Suspension Bridges

Rank	Name	Main Span (m)	Year Opened	Location	Country
1	Akashi Kaikyō Bridge	1991	1998	Kobe-Awaji Island	Japan
2	Xihoumen Bridge	1650	2009	Zhoushan	China
3	Great Belt Bridge	1624	1998	Korsør-Sprogø	Denmark
4	Yi Sun-sin Bridge	1545	2012	Gwangyang-Yeosu	South Korea
5	Runyang Bridge	1490	2005	Yangzhou-Zhenjiang	China
6	Nanjing Fourth Yangtze Bridge	1418	2012	Nanjing	China
7	Humber Bridge	1410	1981	Hessle-Barton-upon-Humber	United Kingdom
8	Jiangyin Bridge	1385	1999	Jiangyin-Jingjiang	China
9	Tsing Ma Bridge	1377	1997	Tsing Yi-Ma Wan	Hong Kong
10	Hardanger Bridge	1310	2013	Vallavik-Bu	Norway

Bridge (Fig. 1.27) crosses the busy Akashi Strait and links the city of Kobe on the mainland of Honshu to Iwaya on Awaji Island, in Japan. Since its completion in 1998, the bridge has had the longest central span of any suspension bridge in the world at 1991 m. The central spans of the top 10 largest suspension bridges are longer than 1300 m, indicating the incomparable

Fig. 1.27 The Akashi Kaikyō Bridge (Japan, the longest bridge since 1998).

spanning capability of this bridge type. The suspension bridge will be discussed in detail in Chapter 11.

1.4 SELECTION OF BRIDGE TYPES

The selection of the proper type of bridge is determined based on the results of topographic survey, geological survey, traffic survey, geotechnical survey, hydro technical survey, seismic survey, and meteorological survey, etc., as well as the cost, environmental impact, and esthetics. While, the maximum span length (or spanning capability) is generally an important factor that should be considered for designing the bridge superstructure due to the fact that each bridge type has its own scope of application. As a matter of experience, the appropriate span length of each structural form is summarized in Table 1.5.

Selection of the bridge superstructures is closely related to the use of construction materials. Based on the materials used for superstructure construction, the modern bridges can be roughly divided into concrete bridges and steel bridges, with different structural forms. Benefit by the high strength to weight ratio, steel construction requires less material than other traditional technologies and contributes to reducing a bridge's environmental impact. The steel bridges are generally built in large spans such as arch bridges, truss bridges, cable-stayed bridges, and suspensions bridges. Especially for large span bridges, as the dead weight governs the load carry capacity of bridges, the bridge superstructures are built in steel but not concrete. Concrete is a brittle material, like stone, good in compression but weak in tension, so it is vulnerable to crack under bending or twist. Concrete has to be reinforced with steel to improve its ductility, naturally its emergence follows the development of steel. However, for some structural forms of bridges, concrete will be a perfect material to build, such as the arch bridges whose members are mainly under compression. Also, concrete bridges are also widely used for short-span bridges due to the relative low cost and less maintenance in service stage. In addition, with the development of the prestressing technique, the prestressed concrete bridges can also be built in medium spans. The availability of the construction materials should be considered in the selection of the bridge superstructures.

The mechanical characteristics of each bridge type are the determinant factor for an appropriate span capacity. Based on the discussion above, the simply supported structure is statically determinate and is simplest to design, and generally is suitable for short-span bridges. When unyielding foundation

Table 1.5 Structural Form of Bridge Superstructure and Appropriate Span Length

	Structural form	Span length
Steel bridges	Simply supported I-girder	20–50
	Continuous I-girder	30–60
	Simply supported box-girder	50–70
	Continuous box-girder	50–100
	Simply supported truss	70–100
	Continuous truss	70–440 m
	Cantilever truss	70–549 m
	Arch bridge	50–552 m
	Cable stayed bridge	60–1104 m
	Suspension bridge	90–1991 m
Concrete bridges	RC simply supported I-girder	10–20
	PC pretension girder	10–20
	PC post-tension girder	10–90

is attainable, the rigid-frame bridges and arch bridges may provide the most economical solution for span length. For medium-span bridges, the continuous girder bridges, the truss bridges, and the arch bridges can be considered as an alternative. For large span bridges longer than 500 m, the cable-stayed bridge and the suspension bridges are promising solutions. A cable-stayed bridge is the successor to the suspension bridge for spans up to 600 m, and the largest span of cable-stayed bridge 1104 m. However, for super bridges with span length larger than 1000 m, a suspension bridge is still the best choice.

The bridge foundation is another factor that may affect the selection of the bridge superstructures. When unyielding foundation is attainable for building the intermediate piers, then continuous girders supported by independent piers and multispan rigid frames will be good options. When unyielding foundation is available for building the abutments, the arch and rigid-frame bridge can be alternatives. For soft foundations, other bridge types with larger spanning capacity should be selected to avoid the intermediate piers.

To sum up, each bridge type has its own suitable range of application and should be considered in the selection of the bridge superstructures. In addition, other factors such as the cost, environment impact, and esthetics need also to be considered to determine suitable alternatives for bridge superstructures, which will be discussed in Chapter 2.

1.5 EXERCISES

1. Classify the bridge's superstructures according to the materials of constructions, span length, interspan relation, deck location, geometric shape span types, usage, and structural forms.
2. Describe the structural characteristic of Girder Bridge, rigid-frame bridge, truss bridge, arch bridge, cable-stayed bridge, and suspension bridge, respectively.
3. Describe the following terminologies: (a) superstructure, (b) substructure, (c) piers, (d) abutments, (e) span length, (f) total length, (g) bridge width, and (h) clearance.
4. List more than three types of moveable bridges, and describe their characteristics.
5. A bridge is planned to be built over a river as shown below. Please propose three preliminary designs (with different structural forms) including

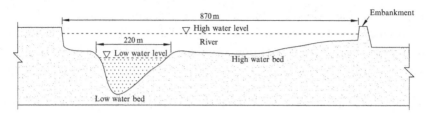

Fig. 1.28 Cross section of a river.

the span numbers and span lengths. However, it should be noted that no bridge piers should be designed in the low water bed zone (Fig. 1.28).

REFERENCES

AASHTO, 2004. AASHTO LRFD Bridge Design Specifications, third ed. with 2005 Interims. American Association of State Highway and Transportation Officials, Washington, DC.

Abrar, M., Masood, W., 2014. Commercial building of reinforced concrete design. Int. J. Eng. Sci. Res. Technol. 3 (5), 890–895.

Catbas, F.N., Grimmelsman, K.A., Barrish, R.A., Tsikos, C.J., Aktan, A.E., 1999. Structural identification and health monitoring of a long span bridge. In: Chang, F.-K. (Ed.), Structural Health Monitoring 2000. CRC Press, Lancaster, Pennsylvania.

Hayashikawa, T., 2000. Bridge Engineering. Asakura Publishing Co. Ltd., Tokyo, Japan.

Khan, M.A., 2010. Bridge and Highway Structure Rehabilitation and Repair. The McGraw-Hill Companies Inc., Columbus, United States.

Lutomirska, M., Nowak, A.S., 2013. Site Specific Live Load and Extreme Live Load Models for Long Span Bridges, Safety, Reliability, Risk and Life-Cycle Performance of Structures and Infrastructures. CRC Press, Leiden, Netherland.

Ponnuswamy, S., 2008. Bridge Engineering, second ed. Tata McGraw-Hill, New Delhi, India.

Portland Cement Association, 1936. Analysis of Rigid Frame Concrete Bridges. Portland Cement Association, Chicago.

Qin, Q., Mei, G., Xu, G., 2013. Chapter 20: bridge engineering in China. In: Chen, W.F., Duan, L. (Eds.), Handbook of International Bridge Engineering. CRC Press, Boca Raton, FL.

Schneider, C.C., 1907. Movable bridges. Proc. Am. Soc. Civ. Eng. 33 (Part 1), 154.

Smith, D.A., Dykman, P.T., Norman, J.B., 1989. Historic Highway Bridges of Oregon. Oregon Historical Society Press, Portland, Oregon.

Taly, N., 1997. Design of Modern Highway Bridges. McGraw-Hill, New York.

Tang, M.-C., 2016. Super-long span bridges. Struct. Infrastruct. Eng., 1–9.

CHAPTER TWO

Bridge Planning and Design

2.1 INTRODUCTION

The bridge structures are important component in highway, railway, and urban road and play important roles in economy, politics, culture, as well as national defense. Especially for medium span and larger span bridges, they are generally served as "lifeline" engineering due to their vital functions in the transportation network. Therefore, the bridge structures should be carefully planned and designed before the construction. The bridge design process, bridge design philosophy will be discussed in this chapter.

A brief diagram showing the bridge planning and design process is shown in Fig. 2.1. In bridge design survey, planning, and design, the structural safety, serviceability, economic efficiency constructability, feasibility in structural maintenance, environmental impact, etc., should be considered to propose an appropriate bridge location and suitable structural type.

2.2 BRIDGE DESIGN PHILOSOPHY

Two thousand years ago, in "De Arhitectura," Marcus Vitruvius Pillo proclaimed: "structures shall be safe, functional and beautiful" (Tang, 2006). Until today, we still cannot escape from the three goals but only modify this slightly to: "A bridge must be safe, functional, economical and beautiful!" Although there are several different semantics and different ways to express concepts of the bridge design philosophy, but essentially the design philosophy for modern bridges are similar among different design codes of different countries. As an example, the bridge design philosophy specified in Japanese bridge design specification is shown below.

"In designing a bridge, the fitness to the purpose of use, safety of structures, durability, securing of the construction quality, reliability and ease of maintenance, environmental compatibility, and economy should be taken into consideration."

JRA, Specification for Highway Bridges

Bridge Engineering
http://dx.doi.org/10.1016/B978-0-12-804432-2.00002-5

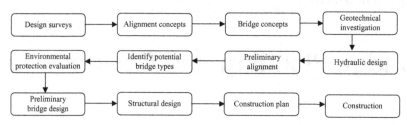

Fig. 2.1 Bridge survey and design process.

According to the explanation of JRA, the fitness to the purpose of use means the bridge's function of being available to traffic as planned, including the serviceability for users to use safely and comfortably. The bridge safety requires that the bridge has enough load carrying capacity to dead loads, live loads, seismic load, etc., that may occur in the bridge service stage.

The durability means that with aging, the performance of a bridge will not suffer significant degradation with respect to the bridge safety and serviceability. Securing of the constructability means the proposed bridge design should be able to be achieved by using the available technology and ensuring the structural safety in both the construction stage and the service stage, as well as the durability.

Reliability and ease of maintenance requires that the repair and member replacement work shall be performed easily when damage or deterioration occurs. Planning suitable maintenance method in the design stage as design preconditions is essential to ensure that various inspections scheduled to be carried out during the in-service stage.

Environmental compatibility means the impact of the bridge construction on features of local environment, such as marine life, wildlife along river banks, riverbed, flora and fauna along river banks, archeological sites, etc., need to be considered.

Then finally, the economy, or economic efficiency, means the life cycle cost of the bridge should be minimized. The life cycle cost means the sums up of all relevant costs of a bridge structure over a given study period not only include the initial cost but also include the maintenance and inspection cost, future rehabilitation costs, and the removal cost.

In essence, a bridge is a civil engineering structure aiming at an efficient balance of loads or forces, from where they are applied to the foundation. The serviceability (without severe cracking or large deformation, etc.) during the bridge life should be able to be guaranteed. The bridge elegance must come from the proportions, the shapes, which have to evidence and express this flow of force, slenderness and transparency will come from a minimal and thus efficient use of constructional materials. Besides, the designers are also responsible for the cost of their constructions because they are mainly paid by public resources. Esthetics and functionality cannot be separated; the designers must achieve architectural perfection in a full respect of functional, financial, and structural need. To achieve these goals, the bridge engineers should always work together with architecture or artist to perfectly express the structural concept.

2.3 BRIDGE SURVEY

Bridge surveying is important because it can provide information for the whole bridge design process. Though reconnaissance surveys are generally made at all possible bridge sites and provide information for bridge location and bridge type selection, a detailed survey is performed at the best suitable site to get information for the bridge design and construction planning.

The bridge survey mainly includes the topographic survey, traffic survey, geological survey, hydrotechnical survey, seismic survey, and meteorological survey. Traffic survey needs to be first conducted for predicting the amount of traffic at various stages during the service life of the bridge and thus demonstrates the necessity and importance of the new bridge. The topographic survey and geological survey are then performed to determine a topographic map and the geologic map, respectively, which can be used for determining the bridge location, structural type, bridge length, as well as the span length ratio. Geotechnical survey, including the soil experiment, underground water level, and hydrotechnical survey investigation on cross-sectional river shapes (in case of building a bridge crossing a river), tide level (in case of building a bridge in a lake), water level, and navigation ships, should be conducted to provide information for design and construction of the bridge foundation. In addition, seismic survey forcing on seismographic record and earthquake disaster records and meteorological survey, investigating on records of wind speed, air temperature, rainfall, and snowfall, should also be performed.

2.4 BRIDGE PLANNING AND GEOMETRIC DESIGN

In bridge planning, a bridge location and structural type should be decided according to the route alignment, topography, geology, meteorology, crossing object, and other external conditions. Geometric design for bridge structures includes the graphic design, horizontal design, vertical design, design of geometric cross sections, intersections, and various design details. The goals of geometric design are to maximize the serviceability, structural safety, economy of facilities, and structural esthetics, while minimizing their environmental impacts.

2.4.1 Horizontal Layout

First, the bridge location should be decided. In general, the culvert and small bridges should be following the route direction of the main road. By considering the hydrology and curves on the main road, the bridges can be designed as curved or skew bridges. For medium and large bridges, however, the bridge location should be determined according to the main route direction if possible, and the overall consideration of both road and bridge is necessary. A straight channel with stable water flow and geological conditions will be selected. In addition, the horizontal curve radius, super elevation and broaden, easement curve, and set-up of the speed-change lanes should be designed according to the design specifications.

2.4.2 Longitudinal Elevation

The bridge horizontal (or longitudinal) design includes the total span length, the number of spans, the bridge elevation and longitudinal slope, the burial depth of the foundation, etc.

2.4.2.1 Total Length

In general, the total length of the bridge should be determined according to the hydrological conditions. In the design life of a bridge, the design flood discharge shall be ensured, and the drift ices, vessels, raft, and other drifting objects in the water should be able to pass through the bridge. Adverse change of the waterway due to the over compression of the riverbed should be avoided. In addition, under some circumstances, it is possible to shorten the bridge length for deep buried foundation, but the river-bed scouring or erosion that may affect the bridge foundations should be carefully checked and avoided in the design.

2.4.2.2 Number of Spans

For a long bridge, the total length is generally divided into several spans. The span numbers, however, will not only affect the esthetic appearance and constructional difficulties but will also influence the total cost of the bridge to a great extent. For example, the larger span length will result in smaller number of span and then reduces the cost of the foundations, but the cost of the superstructure will increase. On the contrary, the larger span numbers and smaller span length will result in relatively low cost of the bridge superstructure and high cost of the substructure. Therefore, appropriate bridge span numbers should be determined on the basis of the most economical design considering both bridge superstructure and substructure. All in all, the determination of the span numbers for medium and large span bridge is a complex problem and should be determined according to the serviceability, bridge location and environment, geological conditions, hydrologic condition, and economic efficiency.

2.4.3 Transverse Cross Section

The bridge cross section is mainly determined according to the bridge width and bridge structural type. Bridge width is designed on the basis of the traffic demand and generally taken as the same as the road width that the bridge located at. The bridge clearance limit (above the deck) is mainly determined by the importance of the bridge and design speed of the highway.

2.5 BRIDGE DESIGN METHODS
2.5.1 Allowable Stress Design

Allowable Stress Design (ASD) is also referred to as the service load design or working stress design (WSD). The basic conception (or design philosophy) of this method is that the maximum stress in a structural member is always smaller than a certain allowable stress in bridge working or service conditions. The allowable stress of a material determined according to its nominal strength over the safety factor. Therefore, the design equation of the ASD method can be expressed as:

$$\sum \sigma_i \le \sigma_{all} = \frac{\sigma_n}{F_s} \tag{2.1}$$

where σ_i is a working stress due to the design load, which is determined by an elastic structural analysis under the design loading conditions. σ_{all} is the allowable stress of the constructional material. The σ_n is the nominal stress of the material, and F_S denotes the safety factor specified in the design specification. Selection of allowable stress depends on several factors, such as the design code, construction materials, stress conditions, etc. Taking the allowable of SS400 (a structural steel in Japanese design code) in tension as an example, the allowable stress shall be taken as 140 MPa when its thickness is larger smaller than 40 mm but 125 MPa for thickness larger than 40 mm. When it is in compression, the buckling may also be considered in selecting the allowable stress.

The ASD method is very simple in use, but it cannot give a true safety factor against failure. All uncertainties in loads and material resistance are considered by using the safety factor in ASD. Although there are some draw-backs to ASD, bridges designed based on ASD have served very well with safety inherent in the system. Currently, ASD design method is still used in the bridge design specifications in Japan.

2.5.2 Load Factor Design

To overcome the drawbacks of the ASD design method, the ultimate load design method was developed in reinforced concrete design, which was modified as the Load Factor Method Design (LFD). In this method, different load multipliers was introduced, and the LFD design equation generally can be expressed as:

$$\sum \gamma_i Q_i \leq \phi R_n \tag{2.2}$$

where γ_i is a load factor and ϕ is the strength reduction factor, Q_i and R_n are, respectively, load effect and nominal resistance.

2.5.3 Load and Resistance Factor Design

Currently, limit state design (LSD) is the most popular design concept for bridge design and widely used for many countries in the world. In the United States, it is known as load and resistance factor design (LRFD). Load and resistance factor design is a design methodology in which applicable fail-ure and serviceability conditions can be evaluated considering the uncer-tainties associated with loads by using load factors and material resistances by considering resistance factors. The LRFD was approved by AASHTO in 1994 in the LRFD Highway Bridge Design Specifications.

$$\sum \eta_i \gamma_i Q_i \leq \phi R_n = R_r \tag{2.3}$$

Eq. (2.3) is the basis of LFRD methodology (AASHTO, 2007). In this equation, η_i is the load modifier, γ_i is the load factor, ϕ is the resistance factor, Q_i and R_n are load effect and nominal resistance, respectively.

Several limit states, including strength limit state, service limit state, the fatigue and fracture limit state, and the extreme event limit state, are included in this design method. The strength and stability are considered in the strength limit state design. In service limit state design, the stress, deformation, and drack width in service condition should be carefully checked. Stress ranges, stress cycles, and toughness requirement are considered in the fatigue and fracture limit state, and the survival of a bridge during a major earthquake or flood is considered in extreme event limit state.

Though the current design specification in Japan is based on the ASD design, the LRFD method is also used for designing the Tokyo Gate Bridge in Japan.

2.6 EARTHQUAKE- AND WIND-RESISTANT DESIGNS

2.6.1 Earthquake Resistant Design

In general, the methods of earthquake-resistant design are different in different countries due to their localities. In this section, earthquake-resistant design in Japan is mainly dealt with (Japan Bridge Association, 2012):

(1) Need to consider near-field ground motions in earthquake-resistant design

The return period of an active fault is thought to be about 1000 years. The likelihood of such a disaster occurring over a period of 50 years is roughly 5%. Since the level of risk is low, strategic judgments must be made in order to maintain the capacity of civil engineering structures to withstand earthquakes. However, there have been quite a few instances in which serious damage has resulted from inland earthquakes with a magnitude of 7 or more. Therefore, even though the risk level is low, it is still possible for strong earthquakes of this type to strike somewhere, so their potential for disaster should not be ignored.

(2) Ground motion in earthquake-resistant design

Two types of earthquake motions should be considered in assessing the aseismic capacity of bridges. The first type is likely to strike a bridge once or twice during its service life. The second type is very unlikely to strike a bridge during the structure's life time, but when it does, it is extremely strong. The second type ground motion includes those generated by interplate earthquakes in the ocean and those generated by earthquakes by inland faults. The concepts behind these two types of motion have been incorporated into the existing earthquake-resistant design of some structures, and these two types of the ground motions are called "Level I earthquake motions" and "Level II earthquake motions." Objectives for and characteristics of these earthquake motions in earthquake-resistant design are as follows (JRA, 2012d):

(a) Level I earthquake motions is the level in which structures are not damaged when these motions strike.

(b) Level II earthquake motions is the level in which an ultimate capacity of earthquake resistance of a structure is assessed in plastic deformation range.

Level I earthquake motions are used in conjunction with the elastic design method and are established as earthquake motions for static load analysis or elastic dynamic analysis. There are many different types of civil engineering structures, and systems of and knowledge about the design methods for each of them have been developed through experience. In existing design systems of road bridges Level II earthquake motions are treated as design earthquake motions with an elastic response of 9.8 m/s^2 on standard ground. A problem specific to direct inland earthquakes is that the relative displacement caused by the dislocation of an earthquake fault reaches the ground surface and structures straddle the fault. Using existing technology to deal with this situation is problematic because of the difficulty of specifying the exact locations of faults and the inevitability in many cases of linear structures crossing faults. Solutions to these problems require further research and development.

(3) Level II earthquake motions

Level II earthquake motions generated by active inland faults are determined based on indentification of active faults that threaten an area and assumptions of source mechanism, through comprehensive examination of geological information on active faults, geodetic information on diastrophism, and seismological information on earthquake activity. Considerable effort must be done in establishing engineering methods.

(4) How Level II earthquake motions are expressed

(a) Level II earthquake motions are basically used for earthquake-resistant design based on damage control concepts. Therefore, the dynamic characteristics of earthquake motions should be expressed concisely, such as in the response spectrum or time history waveforms.

(b) Ground levels where earthquake motions are given

(i) Earthquake motions on bedrock: Basically, Level II earthquake motions are established in bedrock. It was pointed out that the irregularity of the topography in the area greatly affected local amplification effects of the earthquake motions. Furthermore, the nonlinear characteristics of the surface layer and softening of sandy ground greatly affected the amplification characteristics. To specify earthquake motions by evaluating these phenomena, it is essential to examine information on three-dimensional ground structures on bedrock, as well as to accumulate more information on topographic features and ground conditions and to conduct more research and development.

(ii) Earthquake motions on basement from engineering viewpoint (engineering bedrock): Earthquake motions on engineering bedrock are established by back analysis of earthquake motions observed on the ground surfaces.

(iii) Earthquake motions on ground surfaces: There are few records from observations of earthquake motions on bedrock and engineering bedrock, so in many cases earthquake motions on bedrock and engineering bedrock may not be able to be specified. Because of this, for now, earthquake motions are established on ground surfaces for which records of strong earthquake motions exist.

(5) Required aseismic capacity and earthquake-resistant design of superstructures

(a) Earthquake resistance to Level I earthquakes

In principle, no damage should occur to any structure when earthquake motion of Level I occurs. Accordingly, the dynamic response during motion of this level should not exceed the elastic limit.

(b) Earthquake resistance to Level II earthquakes

Important structures and structures requiring immediate restoration in the event of an earthquake should, in principle, be designed to be relatively easily repairable, even if damage is suffered in the inelastic range. Accordingly, the maximum earthquake response of such structures must not exceed the allowable plastic deformation or the limit of ultimate strength. For other structures, complete collapse should not occur even if damage is beyond repair. Accordingly, deformation during an earthquake of this level should not exceed the ultimate deformation.

The degree of importance of structures can be determined based on the following factors:

(i) The effect of structural damage on life and survival

(ii) The effect of structural damage on evacuation, relief, and rescue operations

(iii) The effect of structural damage on everyday functions and economic activities

(c) Important issues in the earthquake-resistant design of superstructures and related topics for research and development in evaluating the dynamic response of a structure to Level I earthquakes, linear multimode response analysis using response spectra or time history earthquake motions is recommended. Further, an investigation of the three-dimensional effects, including vertical motion, should be carried out when necessary.

In evaluating the dynamic response of a structure to Level II earthquakes, elastoplastic time history response analysis is recommended. However, it is also acceptable to use practical and more convenient methods based on equivalent linearization analysis or design spectra corresponding to the allowable ductility factor. For structures with a low degree of static indeterminacy, a rigorous verification of the ability to carry sustained loads is required, especially in the case of a Level II earthquake. Accordingly, it is desirable to investigate the accuracy of various elastoplastic analysis methods and compare them with test results. For any structures with a high degree of static indeterminacy, including steel and concrete structures, an ultimate deformation analysis that takes into account the damage process is recommended.

Since the earthquake response of short-period structures is determined by the effect of dynamic interactions of the foundation-ground system in the nonlinear range, research into design methods that take account of this should be promoted. It may prove possible to use a simplified procedure in which the effect of dynamic interactions is incorporated into seismic design by lengthening the natural period of the total structural system and increasing the damping coefficient.

In order to enhance the earthquake resistance of structures, introduction of new technologies such as seismic isolation and active control is recommended. Seismic isolation increases the deformability and damping capacity of relatively short-period structures, while the use of active control incorporating energy absorbing mechanisms can increase the damping capacity of long-period structures.

(6) Case study

    (a) Hyogoken Nanbu Earthquake (1995)

        Girders were fallen in many bridges due to malfunction of unseating prevention devices associated with the collapse of substructures, as shown in Fig. 2.2. Accordingly, the specifications were revised to performance-based design methodology.

    (b) The Great East Japan Earthquake in 2011

        The 2011 off the pacific coast Tohoku earthquake occurred on the 11th of Mar. 2011. Powerful tsunami waves were caused by this earthquake and destroyed the cities in coastal area of Tohoku region. The main cause of many casualties in the Earthquake was reported as tsunami and about 90% of the casualties were resulted from it.

        In the area whose seismic intensity was large, many bridges without damage could be found. This is due to the continuous amelioration of seismic design criteria in Japan. One of the typical damages in steel bridges could be seen in shoes made from steel (Fig. 2.3). On the other hand, few rubber bearings are found broken. It follows from this that, in order to reduce the damages

Fig. 2.2 The Hyogoken Nanbu Earthquake (1995).

Fig. 2.3 The 2011 off the pacific coast Tohoku earthquake.

Fig. 2.4 Damaged bearing.

shown in Fig. 2.4, a high-damping rubber bearing is recommended as a substitute of the cast steel shoe.

Even if we shall confine damage reduction countermeasures in
the field of steel bridges, there are a lot of countermeasures to

enhance the potential of earthquake-resistant capacity. Rigidly connected girders and piers are examples to make bridge structures earthquake resistant.

(c) Damage due to TSUNAMI

Steel bridge superstructures were swept away as shown in Fig. 2.5. And some bridges became deformed by collision such as floating ships on tsunami (Fig. 2.6).

The current design criteria do not take into account the forces due to tsunamis. On the basis of scientific knowledge such as analyses of ancient documents and surveys of tsunami deposits and coastal topography, necessary revisions of earthquake and tsunami countermeasures are expected (Cabinet Office, 2007/2012).

Fall-off prevention devices for steel bridges tore off by tsunami waves. The wave force of tsunami could easily tear off the connectors of fall-off prevention devices which are strong enough to carry the dead load of the bridge itself, as shown in Fig. 2.7. Therefore, the fall-off prevention devices are almost useless to prevent the bridge from being washed away by the large tsunami waves (Kasano et al., 2012).

In view of the height of bridge superstructures, the bridge superstructures that were located in lower height from the ground level tend to withstand the tsunami waves. On the contrary, the

Fig. 2.5 Bridge partially swept away.

Fig. 2.6 Ship collision damage.

Fig. 2.7 Tearing off of fall-off prevention devices.

bridge superstructures on the high piers are likely to be washed away when subjected to the tsunami waves (Figs. 2.8 and 2.9).

The phenomenon that the bridge superstructures with high clearance tends to be swept away by tsunami waves is opposite to the result brought by the equation currently used to estimate

Fig. 2.8 Shin-kitakami oohashi.

Fig. 2.9 Truss members being swept away.

the wave force in Japan. Therefore, in order to confirm this phenomenon, the tsunami wave forces acting on a bridge deck should be examined. Some fundamental studies on the tsunami wave forces acting on a bridge have started since the 2004 Indian Ocean earthquake and tsunami (Iemura et al., 2007; Murakami et al., 2009; Kataoka et al., 2006; Shoji et al., 2009). Generally the removal of a bridge deck is caused by the lateral wave force in conjunction with vertical force (Araki et al., 2010). However the detailed flow mechanism of tsunami waves has not been explained yet. Furthermore, tsunami waves also carried debris including ships on their surface. Then, it is expected that the debris collided with a bridge superstructure and pushed it away. Therefore, if the bridge deck is located as high as the height of tsunami waves, the deck is subjected to not only the wave force but also the impact force of crashing debris (Iemura et al., 2005).

2.6.2 Wind-Resistant Design

(1) General

In the first half of the 19th century, suspension bridges occasionally collapsed under wind loads because girders tended to have insufficient rigidity. In the latter half of the 19th century, such collapses decreased because the importance of making girders sufficiently stiff was recognized.

In the beginning of the 20th century, stiffening girders with less rigidity reappeared as the deflection theory was applied to long-span suspension bridges. The Tacoma Narrows Bridge collapsed 4 months after its completion in 1940 under a wind velocity of only 19 m/s. The deck of the bridge was stiffened with I-girders formed from built-up plates. I-girders had low rigidity and aerodynamic stability was very inferior as shown in recent wind-resistant design. After this accident, wind tunnel tests for stiffening girders became routine in the investigation of aerodynamic stability. Truss-type stiffening girders, which give sufficient rigidity and combined partially with open deck grating, have dominated the design of modern suspension bridges in the United States. A new type of stiffening girder, however, a streamlined box girder with sufficient aerodynamic stability, was adopted for the Severn Bridge in the United Kingdom in 1966. In the 1980s, it was confirmed that a box girder, with big fairings (stabilizers) on each side and longitudinal openings on upper and lower decks, had excellent aerodynamic stability. This concept was adopted for the Tsing Ma Bridge, completed in 1997. The Akashi Kaikyo Bridge has a vertical stabilizer in the center span located along the centerline of the truss-type stiffening girder just below the deck to improve aerodynamic stability.

In the 1990s, in Italy, a new girder type has been proposed for the Messina Straits Bridge, which would have a center span of 3300 m. The 60-m-wide girder would be made up of three oval box girders which support the highway and railway traffic. Aerodynamic dampers combined with wind screens would also be installed at both edges of the girder.

(2) Design standard

Fig. 2.10 shows the wind-resistant design procedure specified in the Honshu-Shikoku Bridge Standard (Harazaki et al, 2000). In the design procedure, wind tunnel testing is required for two purposes: one is to verify the airflow drag, lift, and moment coefficients which strongly influence the static design, and the other is to verify that harmful vibrations would not occur.

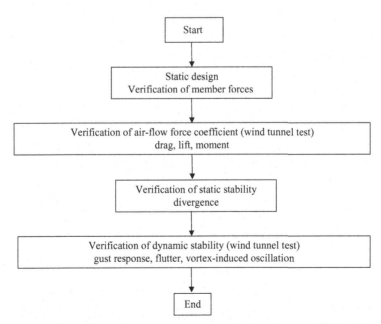

Fig. 2.10 Procedure for wind-resistant design. *(Source: Based on Honshu-Shikoku Bridge Authority, Wind-Resistant Design Standard for the Akashi Kaikyo Bridge, HSBA, Japan, 1990.)*

(3) Analysis

Gust response analysis is an analytical method to ascertain the forced vibration of the structure by wind gusts. The results are used to calculate structural deformations and stress in addition to those caused by mean wind. Divergence, one type of static instability, is analyzed by using finite displacement analysis to examine the relationship between wind force and deformation. Flutter is the most critical phenomenon in considering the dynamic stability of suspension bridges because of the possibility of collapse. Flutter analysis usually involves solving the motion equation of the bridge as a complex eigenvalue problem where unsteady aerodynamic forces from wind tunnel tests are applied.

(4) Wind tunnel testing

In general, the following wind tunnel tests are conducted to investigate the aerodynamic stability of the stiffening girder.

(a) Two-Dimensional Test of Rigid Model with Spring Support: The aerodynamic characteristics of a specific mode can be studied. The scale of the model is generally higher than 1/100 of the prototype.

(b) Three-Dimensional Global Model Test: Test used to examine the coupling effects of different modes.

For the Akashi Kaikyo Bridge, a global 1/100 model about 40 m in total length was tested in a boundary layer wind tunnel laboratory. Together with the verification of the aerodynamic stability of the Akashi Kaikyo Bridge, new findings in flutter analysis and gust response analysis were established from the test results.

(5) Countermeasures against vibration

Countermeasures against vibration due to wind are classified as

(a) Increase structural damping: Damping, a countermeasure based on structural mechanics, is effective in decreasing the amplitude of vortex-induced oscillations which are often observed during the construction of the main towers and so on. Tuned mass dampers (TMD) and tuned liquid dampers (TLD) have also been used to counter this phenomenon in recent years. Active mass dampers (AMD), which can suppress vibration amplitudes over a wider frequency band, have also been introduced.

(b) Increase rigidity: One way to increase rigidity is to increase the girder height. This is an effective measure for suppressing flutter.

(c) Aerodynamic mechanics: It may also be necessary to adopt aerodynamic countermeasures, such as providing openings in the deck, and supplements for stabilization in the stiffening girder.

2.7 BRIDGE DESIGN SPECIFICATIONS

Bridge design specifications, or design standard (or code) is established to ensure bridge safety in terms of stiffness, strength, and stability of the whole bridge or each bridge component. There are many bridge design specifications in the world, and the followings are given as examples.

2.7.1 Bridge Specifications in the United States

The first national design speciation for highway bridges in the United States was published by the American Association of State Highway Officials (AASHO) in 1927. The ASD had been used until 1970 before the new design method of load factor design (LFD) was development. Probability-based load and resistance factor design (LRFD) philosophy was first approved in 1994 in AASHTO LRFD Bridge Design Specification, but it was not widely used until 2003. AASHTO specifications for highway bridges have been revised several times based on technical progress and changes of social needs.

2.7.2 Bridge Specifications in Japan

The "Specifications for Highway Bridges" is the most basic highway bridge design code in Japan. It applies to the bridges with the span length <200 m. Japanese Design Specifications for Highway Bridges (JRA, 2012a,b,c,d) consist of five parts: Part I Common, Part II Steel Bridges, Part III Concrete Bridges, Part IV Substructures, and Part V Seismic Design. These specifications have been revised for several times, and the latest version was updated in 2012. After the 1995 Kobe Earthquake, the 1996 specifications were revised a lot to enhance seismic design. In the specifications updated at 2002, the concept of performance-based design was introduced (Fukui et al., 2005). Additionally, the design considerations for durability were enhanced so as to design the sustainable structures (Kuwabara et al., 2013; Unjoh et al., 2002). For designing the railway bridges in Japan, Standard specification for railway structures (Railway Technical Research Institute (RTRI), 2009) should be referred to.

2.7.3 Bridge Specifications in United Kingdom

BS 5400 is a British Standard code for the design and construction of high-way, railway and pedestrian bridges made of steel, concrete and composite materials. BS 5400 is made up of several parts (BS 5400-1–10), including the general statement, specification for loads, and code of practice for design of steel bridges, concrete bridges, and composite bridges. In Mar. 2010, the British structural design standards including British bridge code of BS 5400 was withdrawn and superseded by the Eurocodes.

2.7.4 Bridge Specifications in EU

Eurocodes are the suites of European Standards developed by the European Committee for the structural design of civil engineering structures within the European Union. In European Union, it is the responsibility of each national standards body to implement each Eurocode part as a national standard, and all "conflicting national standards" should be withdrawn by 2010 (Hendy and Johnson, 2006). The Eurocodes that applicable to bridge structures include

Eurocode 0: Basis of structural design
Eurocode 1: Actions on structures
Eurocode 2: Design of concrete structures
Eurocode 3: Design of steel structures
Eurocode 4: Design of composite steel and concrete structures
Eurocode 5: Design of timber structures
Eurocode 6: Design of masonry structures
Eurocode 7: Geotechnical design
Eurocode 8: Design of structures for earthquake resistance
Eurocode 9: Design of aluminum structures

2.7.5 Bridge Specifications in China

Two series of bridge design specifications are used in China, including design specifications for highway bridges and design specifications for railway bridges. Six parts are included in the design specifications for highway bridges in 1989. In the specifications, both load and resistance factor design (LRFD) theory for reinforced prestressed concrete members and ASD theory for steel and timber members are adopted. (Li and Xiao, 2000). In 1999, the national standard for Reliability Design of Highway Engineering Structures was published in China. These codes have also been revised for several times, and the LFRD design code was also adopted in the design of Ground Base and Foundation of Highway Bridges and Culvert. (Qin et al., 2013).

2.8 STRUCTURAL DESIGN AND DESIGN DRAWINGS

2.8.1 Structural Design

The structural design for a modern bridge should be conducted according to the existing local design codes, based on theoretically valid method, experimentally verified method, or other appreciated knowledge.

In a bridge design, some factors need to be taken into consideration include: (a) the probability of bridge structural members that may result in local or entire collapse; (b) inspection facilities that are required to conduct in-service inspections, or in the event of accidents; and (c) possible maintenance method for bridge members. For members that are likely to be replaced in the bridge service stage shall be carefully planned in advance to ensure the reliability and ease of maintenance.

2.8.2 Design Drawings

The design drawing refers to papers and other materials prepared for providing reference information (such as the construction condition, and other matters that related to manufacturing and construction) for structural analysis, bridge construction, and maintenance during the service stage. For highway bridges, various temporary members are often installed during the construction, thus lack of those information may lead to inappropriate responses during inspection and maintenance work. The minimum information that should be provided in a design drawing specified by JRA is shown in Fig. 2.11.

2.9 BRIDGE ESTHETIC DESIGN AND A CASE STUDY

2.9.1 Esthetic Design

As mentioned before, a good bridge design should satisfy both functionality and esthetics. Visual design elements of bridges are line (one-dimensional), shape (two-dimensional), form (three-dimensional), color, and texture. The appropriate arrangement of the visual design elements leads to the order, proportion, rhythm, harmony, balance, contrast, scale, illusion, unity, etc.

From the historical view point of esthetic bridge design, the requirements for an esthetic bridge design are (Taly, 1997):

(1) Selection of the most artistic form in consideration of economy

(2) Expressiveness

(3) Symmetry

(1) Route name and bridge location

(2) Bridge name

(3) Responsible engineer

(4) Date of design

(5) Major design conditions

    (A) Bridge classification

    (B) Design outline

    (C) Load condition

    (D) Topographical, geological and ground conditions

    (E) Material condition

    (F) Manufacturing and construction conditions

    (G) Maintenance conditions

    (H) Other relevant matters

Fig. 2.11 Design drawing examples (Japan Road Association, 2002a,b,c,d).

(4) Simplicity and continuity
(5) Harmony and contrast
(6) Scale and proportion
(7) Order and balance
(8) Conformity with environment
(9) Proper combination of materials

Esthetic design objectives should be considered throughout the design process in which designers should concentrate on developing the best design solution for the aforementioned design requirements.

2.9.2 Case Study of Esthetic Design Proposal of a Footbridge

(1) Planning and design concept

The design theme of the new footbridge is determined as "Modern beauty in the ancient city of NARA" by the designers (Lin et al., 2015). There are more than 1000 cherry trees along the river to form beautiful scenery. Based on this, three things must be considered to succeed in designing this bridge: a beautiful piece of architecture, a valuable new transport link, and harmony with surrounding environment.

Fig. 2.12 Image of footbridge.

The image of the footbridge is shown in Fig. 2.12. The bridge, at 26 m in span, has twin steel arches (acting as upper-superstructure) which are inclined like butterfly wings to produce a feeling of containment on the deck and, simultaneously, an openness to the sky. On the other hand, the cable system below the deck is designed as the lower superstructure for both supporting the deck system with the aid of the compression members and balancing the horizontal force induced from the arches. As a whole, the bridge looks like the transformed "ray fish," with incomparable load carrying capacity as well as sleek appearance with special purity. The footbridge across the river was designed to encourage visitors to the bridge and neighboring communities and as a new landmark for this ancient capital city.

(2) Structural characteristics

A new footbridge structure type, named as "self-anchored extroversion arch-cable system," was created for the design of the target bridge, as shown in Fig. 2.13. The bridge features two extroversion arches

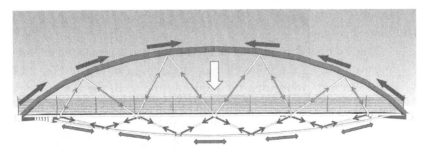

Fig. 2.13 Load transfer paths in the footbridge.

together with the bottom cable systems connecting on the end of the bridge. Hangers (or tie rods) and rigid members were used to transfer external loads from the deck to arch and cable in the form of compression and tension, respectively. The upper arch and the lower cable are connected at both ends of the deck. Therefore, the outward-directed horizontal forces of the arch are borne as tension by the bottom cables, rather than by the ground or the bridge foundations. The deck was designed as steel–concrete composite for enhancing visual effect (or elegant, by reducing the deck thickness), increasing the stability of the bridge in the vertical direction and reducing the structural noise in the service condition. The hangers and rigid bracings were designed as inclined for the purpose of increasing the stability. The bottom cables are effective in increasing the vertical stiffness, while the extroversion arches are capable of enhancing the lateral stiffness of the footbridge. Under the external loads, the arch and the rigid bracings are in compression, while the hangers and the bottom cables are in tension. As all the main load carrying members are mainly in either tension or compression, the proposed design is highly efficient.

(3) Design features

High redundancy: two alternative loading carrying systems (upper arch and lower cable) to ensure the safety of the bridge after failure of one critical member; *high efficiency*: all load carrying members are mainly subjected to axial force; *environmental coordination*: perfect fusion with the surrounding environment; *constructability*: easy construction method and erection procedure; *feasibility in maintenance*: all the structural members can be easily repaired or replaced without affecting the bridge safety and serviceability; *modern elements*: phonetic symbols on the deck surface, external prestressing technique, self-anchored system, etc.; last, but most important is *esthetics*: visual impact due to the extroversion tied arch as well as the lower cable system.

(4) Numerical study

The numerical modeling of footbridge was carried out in three dimensions using the finite-element method, as shown in Fig. 2.14. Beam elements were used for modeling the arch, the bottom rigid bracing, and the transverse beam; the truss elements were used to simulate the hangers and the bottom main cable; solid elements were used to simulate the concrete; and shell elements were employed to model the steel plate. Rebar elements were used for modeling the reinforcement.

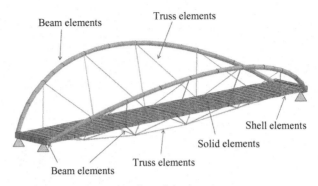

Fig. 2.14 Three-dimensional numerical model.

(5) Relationship between the new bridge and its surroundings

This bridge is designed as a new creative structural type: "self-anchored extroversion arch-cable system." The bridge has twin inclined steel arches, like butterfly wings to produce a feeling of containment on the deck and an openness to the sky. Viewing from the front, a welcoming gesture is created by the inclined arches connecting with the residential area. Viewing from the side, cherry trees on both sides and space above the river way leave the visual impression that the footbridge is floating in the surrounding environment. The perfect combination of the upper arch and lower cable creates a "leaf" or "eye" image for pedestrian traffic, parents and children enjoying the scenery of the river. On the other hand, modern and fashionable handrail, and the phonetic symbols used on the deck surface intensify the modern air of the footbridge. Extroversion arches and handrail also enhance the openness of the footbridge while providing the sense of security (Fig. 2.15).

The dignified and magnanimous appearance of the footbridge creates a lively and positive culture ambience for local people. In the ancient city of Nara, the new footbridge incorporates the latest available technique and constructional materials balanced with an esthetic and graceful appearance that is in perfect harmony with the surrounding environment and local communities.

2.10 EXERCISES

1. Describe the bridge planning and design process, and design considerations.
2. Describe the bridge design philosophy.

Fig. 2.15 Environmental coordination.

3. Describe the bridge design method, and list more than five bridge design specifications (including the one used in your own country) in the world.

4. What is the life cycle cost of a bridge?

REFERENCES

AASHTO, 2007. AASHTO LRFD Bridge Design Specifications. AASHTO, Washington, DC.

Araki, S., Sakashita, Y., Deguchi, I., 2010. Characteristics of horizontal and vertical tsunami wave force acting on bridge beam. J. Coast. Eng. ASCE 66, 796–800.

BS 5400-1, 1988. Steel, concrete and composite bridges. General statement.

BS 5400-2, 2006. Steel, concrete and composite bridges. Specification for loads.

BS 5400-3, 2000. Steel, concrete and composite bridges. Code of practice for design of steel bridges. (This part of standard is being partially replaced).

BS 5400-4, 1990. Steel, concrete and composite bridges. Code of practice for design of concrete bridges.

BS 5400-5:2, 2005. Steel, concrete and composite bridges. Code of practice for design of composite bridges.

BS 5400-6, 1999. Steel, concrete and composite bridges. Specification for materials and workmanship, steel.

BS 5400-7, 1978. Steel, concrete and composite bridges. Specification for materials and workmanship, concrete, reinforcement and prestressing tendons.

BS 5400-8, 1978. Steel, concrete and composite bridges. Recommendations for materials and workmanship, concrete, reinforcement and prestressing tendons.

BS 5400-9.1, 1983. Steel, concrete and composite bridges. Bridge bearings. Code of practice for design of bridge bearings.

BS 5400-9.2, 1983. Steel, concrete and composite bridges. Bridge bearings. Specification for materials, manufacture and installation of bridge bearings.

BS 5400-10, 1999. Steel, concrete and composite bridges. Charts for classification of details for fatigue

Cabinet Office, 2007/2012. Disaster Management in Japan. Government of Japan.

Fukui, J., Shirato, M., Matsui, K., 2005. Performance-Based Specifications for Japanese Highway Bridges. Millpress, Rotterdam.

Harazaki, I., Suzuki, S., Okukawa, A., 2000. Suspension Bridges. In: Chen, W.-F., Duan, L. (Eds.), Bridge Engineering Handbook. CRC Press, Boca Raton, FL.

Hendy, C.R., Johnson, R.P., 2006. Designers' Guide to EN 1994-2 Eurocode 4: Design of Composite and Concrete Structures. Part 2: General Rules and Rules for Bridges. Thomas Telford Publishing, London, United Kingdom.

Iemura, H., Pradono, H.M., Takahashi, Y., 2005. Report on the tsunami damage of bridges in Banda Acer and some possible countermeasures. In: Proceeding of 28th JSCE Earthquake Engineering Symposium, pp. 1–10.

Iemura, H., Pradono, H.M., Yasuda, T., Tada, T., 2007. Experiments of tsunami force acting on bridge models. J. Earthq. Eng. ASCE 29, 902–911.

Japan Bridge Association, 2012. Chronology of Seismic Design Criteria and Survey Report of the Great East Japan Earthquake. JBA, Tokyo.

Japan Road Association, 2012a. Design Specification for Highway Bridges and Commentary, Part I: Common Part, Part II: Steel Bridges. Japan Road Association, Tokyo (in Japanese).

Japan Road Association, 2012b. Design Specification for Highway Bridges and Commentary, Part I: Common Part, Part III: Concrete Bridges. Japan Road Association, Tokyo (in Japanese).

Japan Road Association, 2012c. Design Specification for Highway Bridges and Commentary, Part I: Common Part, Part IV: Substructures. Japan Road Association, Tokyo (in Japanese).

Japan Road Association, 2012d. Design Specification for Highway Bridges and Commentary, Part I: Common Part, Part V: Seismic Design. Japan Road Association, Tokyo (in Japanese).

Kasano, H., Oka, J., Sakurai, J., Kodama, N., Yoda, T., 2012. Investigative research on bridges subjected to tsunami disaster. In: Proceeding of the Australasian Structural Engineering Conference, Perth, Australia in 2011 Off the Pacific Coast of Tohoku Earthquake.

Kataoka, S., Kusakabe, T., Nagaya, K., 2006. Wave force acts on a bridge girder by tsunami. In: Proceeding of the 12th Japan Earthquake Engineering Symposium, pp. 154–157.

Kuwabara, T., Tamakoshi, T., Murakoshi, J., Kimura, Y., Nanazawa, T., Hoshikuma, J., 2013. Outline of Japanese design specifications for highway bridges in 2012. In: Proceedings of the 44st Joint Meeting of U.S.-Japan Panel on Wind and Seismic Effects, UJNR, 2013.

Li, G., Xiao, R., 2000. Chapter 63: bridge design practice in China. In: Chen, W.F., Duan, L. (Eds.), Bridge Engineering Handbook. CRC Press, Boca Raton, FL.

Lin, W., Gui, C., Zheng, S., Lam, H., 2015. Workshop on Elegance in Structures. IABSE NARA.

Murakami, S., Bui, H.H., Nakano, H., Izumo, K., 2009. SPH analysis on tsunami flow around bridge girder. J. Struct. Mech. Earthq. Eng. ASCE 25, 914–919.

Qin, Q., Mei, G., Xu, G., 2013. Chapter 20: bridge engineering in China. In: Chen, W.F., Duan, L. (Eds.), Handbook of International Bridge Engineering. CRC Press, Boca Raton, FL.

Railway Technical Research Institute (RTRI), 2009. Standard specification for railway structures. .

Shoji, G., Moriyama, T., Hiraki, Y., Fujima, K., Sigihara, Y., Kasahara, K., 2009. Evaluation of tsunami wave load acting on a bridge deck subjected to plunging breaker bores and surging breaker bores. J. Coast. Eng. ASCE 65, 826–830.

Taly, N., 1997. Design of Modern Highway Bridges. McGraw-Hill, New York, United States.

Tang, M.-C., 2006. Bridge forms and aesthetics. In: Bridge Maintenance Safety Management and Life-Cycle Optimization. Taylor and Francis Group, London.

Unjoh, S., Nakatani, S., Tamura, K., Fukui, J., Hoshikuma, J., 2002. 2002 seismic design specifications for highway bridges. In: Proceedings of the 34th Joint Meeting of U.S.-Japan Panel on Wind and Seismic Effects, UJNR.

CHAPTER THREE

Materials for Bridge Constructions

3.1 INTRODUCTION

The traditional building materials for bridges are stones, timber, steel, and concrete. Timber and stone can be considered natural materials as they are directly obtained from nature and are used unprocessed in the first period. Bricks were also used but generally included as a subgroup of the stone bridges. After a long period of time, many of the stone bridges are still preserved due to the durable material. As a contrast, few timber bridges still served because their material degrades easily and rapidly due to fire, water, flood, etc. In this period, bridges develop mainly on the basis of the materials but not bridge technology because the materials (such as stones and wood) played an important role in the bridge configuration. The other two, steel and concrete, are artificial, as the raw materials taken from nature need more or less complex processing that changes their physical properties (Troyano, 2003). The appearance of iron and concrete bridges can be considered as the second period and also is the beginning of the modern bridge engineering. In most cases, modern bridges mean the bridges built with steel and concrete, or other new materials. In addition, composite materials formed by fibers of high strength materials are also being used nowadays, not only for new bridges but also for strengthening the existing bridges.

3.2 STONE

Stone bridges have been used in one form for a long history, especially in the form of arches due to their high compressive strength. With stone, the engineers can build both beautiful and durable bridges.

In the history, the Romans were the greatest stone bridge builders. The Romans had a great understanding of loads, geometry, and the material properties of stone, which allowed them to build spans significantly larger than any bridge before their times. At almost the same time, the Chinese were also developing stone bridges such as the very famous Zhaozhou

Bridge Engineering
http://dx.doi.org/10.1016/B978-0-12-804432-2.00003-7

Fig. 3.1 The Nihonbashi bridge. *(Photo by Lin.)*

Bridge. This bridge is the oldest open-spandrel, stone, segmental arch bridge in the world. One of the most famous stone bridges in Japan is the current Nihonbashi (or literally Japan Bridge) in Tokyo constructed in 1911, as shown in Fig. 3.1. Stone bridges are generally built in near the mountains for the easy use of local materials for reducing the cost. But over a long period, well-designed and well-built stone bridges might still the cheapest because they are very long-lasting and need little maintenance during their service lives.

3.3 WOOD OR TIMBERS

Wood or timbers, mainly used for building structures in nowadays, have been an important construction material for bridge structures in the past. Due to the wide range applications of steel and concrete, they are rarely used in bridge construction practice nowadays. However, as the development of chemical techniques for preservation of wood materials, the wooden and timber bridges are again widely used in the engineering practice.

Wood has many advantages as an engineering material. Such as, it has high toughness, and it is an environmentally friendly and a renewable material. Wood has a very high specific strength due to its low density and reasonable strength, and wood's low density also makes it easier to transport. However wood also has some disadvantages like that it is highly anisotropic, highly combustible, and susceptible to termites, woodworm, and infestations. Also, wood cannot be used at high temperatures and is susceptible to rot and disease (University of Cambridge, 2016). There are many timber

Fig. 3.2 The Mathematical Bridge in Cambridge, the United Kingdom. *(Photo by Lin.)*

Fig. 3.3 The Togetsu-kyo Bridge in Kyoto, Japan. *(Photo by Lam.)*

bridges in the world, such as the Mathematical Bridge in Cambridge and the Togetsu-kyo Bridge over the Katsura River in Kyoto, as shown in Figs. 3.2 and 3.3.

3.4 STEEL

In comparison with other bridge materials, steel has very high strength and it is therefore suitable for the construction of bridges with the long spans. Steels are alloys of iron and other elements, primarily carbon. The steel properties, such as the hardness, ductility, and tensile strength, can change a lot according to the amount of alloying elements.

Normal construction steel has both compressive and tensile strengths of several hundred MPa, nearly 10 times the compressive strength of a normal concrete and a hundred times its tensile strength. A prominent advantage of steel is its ductility due to which it deforms remarkably before it breaks, which is also referred to as ductility.

The former Danjobashi Bridge built in 1878 was the first iron bridge produced in Japan, as shown in Fig. 3.4. The Danjobashi Bridge was moved

Fig. 3.4 The Hachimanbashi Bridge (former Danjobashi Bridge). *(Photo by Lin.)*

to its current location and renamed to the Hachimanbashi Bridge in 1929. In 1977, it was designated as a Japanese Important Cultural Property due to its great technical and historical value as a modern bridge. It received the Honor Award from the American Society of Civil Engineers in 1989.

Mechanical properties of structural steel are determined according to the tensile test, in which a steel sample is subjected to a controlled tension until failure, as shown in Fig. 3.5. The conventional stress–strain diagram for steel is shown in Fig. 3.6. In the initial loading stage, the stress–strain curve is a straight line throughout most of the region. The stress is proportional to the strain until the stress limit called the proportional limit, and then the curve tends to bend until the stress reaches the elastic limit. Before this point, the deformation of the steel can return back if the load is removed. A slight increase in stress above this point will result in permanent deformation (plastic deformation), and this behavior is called yielding. The upper yielding point occurs first, followed by lower yielding point, and corresponding stresses are called upper yield stress and lower yield stress, respectively. In engineering, however, the yield stress is generally referred to the lower yield stress. When yielding has ended, an increase in load can be supported by the steel, and the rise of stress in this manner is called strain hardening. As the steel specimen elongates, its cross-sectional area will decrease almost uniformly until the ultimate strength. Thereafter, however, the cross-sectional area will begin to decrease in a localized region of the specimen, which is called necking. The specimen will finally fracture, and the corresponding stress is called fracture strength, as shown in Fig. 3.6.

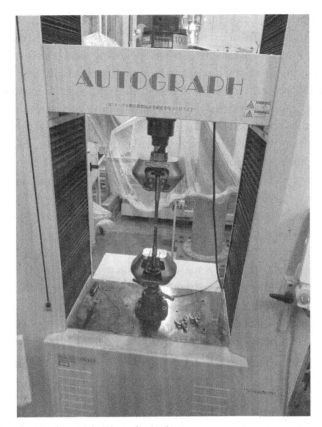

Fig. 3.5 Tensile test for steel. *(Photo by Lin.)*

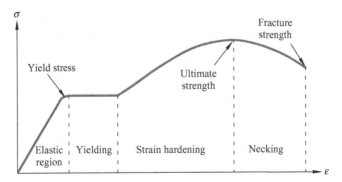

Fig. 3.6 Conventional stress-strain diagrams for steel.

The properties of structural steel are determined by both its chemical composition and manufacture method, and these properties that need to be considered by bridge designers when specifying steel construction products are including strength, ductility, toughness, weld ability, and durability. Steel strength generally denotes the yield strength and the tensile strength (or ultimate strength), and the yield strength is more important in the design as the structures are generally design in the elastic stage. Yield strength is the most common property that the designer will need as it is the basis used for most of the rules given in design codes. While, Japanese design code is an exception, e.g., SS400 denotes the structural steel with a ultimate strength of 400 MPa.

Ductility represents the capacity of steel to deform permanently in response to stress, and the designers rely on ductility for several aspects of design, such as the stress redistribution at the ultimate limit state and bolt group design. A further important property is that of corrosion prevention, such as the use of weathering steel. Other mechanical properties of structural steel that are important to the designer include: Modulus of elasticity, $E = 2.1 \times 10^5$ N/mm^2 (or 2.0×10^5 N/mm^2); shear modulus, $G = E/[2(1+\nu)]$ N/mm^2, often taken as 81,000 N/mm^2 (or 7.7×10^5 N/mm^2 if modulus of elasticity is taken as 2.0×10^5 N/mm^2); Poisson's ratio, $\nu = 0.3$; and coefficient of thermal expansion, $\alpha = 12 \times 10^{-6}/°C$.

3.5 CONCRETE

Concrete is a construction material widely used for bridge construction. Concrete has relatively high compressive strength, but very low tensile strength. For this reason it is usually reinforced with materials, like steel, that are strong in tension. The Young's modulus of concrete is nearly constant at low stress levels but starts decreasing at higher stress levels with the development of cracks. Concrete has a very low coefficient of thermal expansion and shrinks as it matures. All concrete structures crack to some extent, due to shrinkage and tension. Concrete that is subjected to long-duration forces is prone to creep.

Mechanical properties of concrete are determined according to the compressive test, as shown in Fig. 3.7.

Bridge structures are not built with plain concrete but built with reinforced concrete or prestressed concrete. In reinforced concrete (RC, for short), reinforcement is used to improve the concrete's low tensile

Fig. 3.7 Compression test for concrete. *(Photo by Lin.)*

strength as well as its ductility. In modern concrete bridge construction, different reinforced concrete are developed by using steel, polymers, or other composite materials. For a strong, ductile, and durable construction the reinforcement needs to have the several advantages including higher strength, higher tensile strain, thermal compatibility, and durability. In comparison with reinforced concrete, the prestressed concrete bridges are more widely used. In prestressed concrete, compressive stresses are induced by high-strength steel tendons before external loading are applied. Such compressive stress will be used to resist the tensile stresses developed in service stage. Prestress can be induced in prestressed concrete either by pretensioning or by posttensioning the steel reinforcement. Prestressing removes a number of design limitations conventional concrete and allows engineers to design and build lighter and shallower concrete structures without sacrificing strength.

The details of concrete bridges will be discussed in Chapter 6.

3.6 NEW COMPOSITE MATERIALS

Many new materials have been developed in recent years, for both constructing new bridges or repair or strengthening of aged bridges. Fiber-reinforced plastic (FRP) is a typical material of such type, which is composite material made of a polymer matrix reinforced with fibers like glass or carbon. Those materials general have high strength, light weight, high ductility, and flexure capacity. The use of new materials is considered to be a very promising solution to overcome the deterioration problems associated with steel structures because of fatigue or corrosion, and concrete bridges due to the corrosion of reinforcements.

3.7 CASE STUDY—A FAMOUS TIMBER BRIDGE IN JAPAN AND ITS ASSESSMENT

The Kintaikyo Bridge over the Nishiki River at Iwakuni in Japan was originally constructed in 1673. The five-span wooden arch bridge is one of the most historically significant bridges not only in Japan but also in the world. Each span of the three central arches is 35.10 m, the total length being 193.3 m, and the roadway being 5.0 m wide. Each arch of the bridge consists of smoothly curved skeleton lines; its end supports restrained so that when vertical load is applied to that curved surface of an arch, a horizontal reaction force is generated in the supports. When subjected to free vibration, each arch prominently shows the asymmetric mode of deformation inherent to an arch structure, as well as the symmetric mode of deformation. In view of these characteristics, each of the central three spans is considered to have an arch structure. As the floor follows the curve of the arches, the bridge is available only for a pedestrian bridge.

Despite the bridge's unique five-span arch structure, which is designed to enhance durability, the Kintaikyo Bridge, which is primarily made of wood, is vulnerable to natural disaster. As a way of long sustaining the Kintaikyo Bridge, the City of Iwakuni decided to establish a unique system: instead of reinforcing the existing Bridge structure, the City decided to guarantee the succession of Bridge building technology, so as to ensure repeated rebuilding of the Bridge. This solution is indeed unprecedented and unique in the history of bridges in the wood (Figs. 3.8 and 3.9).

Fig. 3.8 Kintaikyo Bridge as of 1998.

Fig. 3.9 Side view of an arch.

The objectives of the assessment were to estimate the present load carrying capacity, to identify any structural deficiencies in the original design and to determine reasons for existing problems identified by the inspection (Yoda, 1998/2004). Load-carrying capacity is an important aspect affecting the safety of the bridge. Pedestrian bridges are no exceptions. Information regarding the ultimate strength of the bridge is required for appropriate allocation of bridge maintenance funds. Measurements of the response to static loading may be used to measure the elastic response of the bridge. However, this type of test requires significant extrapolation of the measurements, if used to predict the strength at design load level. The load carrying capacity here does not refer to the ultimate capacity, rather a lower level referred to as serviceability level.

Four load cases were used to maximize load effects in respective arches. The 112 students weighing 536 N in an average (total weight: 60 kN) were positioned transversely four or two lines to the respective guide rails to balance the load effects in the arch ribs (Fig. 3.10). Structural responses were recorded immediately before and after students were moved. No tourists were allowed on the bridge during testing. The load tests were limited to recording displacements. Instrumentation involved the use of strain gages to determine displacements under static and dynamic loading. The only type of sensors used was resistance strain gages. A total of 18 strain gages were installed within a three span section located mid-span and quarter-span during the field-testing program.

The maximum quarter-span displacement recorded, with static loading of semiuniformly distributed load, was approximately 2.5 mm in the Center Arch as of 1998. And the displacements of both upstream and downstream sides are reasonably close to each other. It follows from this that the effect of load distribution is satisfactory ensuring the soundness of the arch ribs. It indicates that the bridge is expected to be safe at least up to the design pedestrian load (600 kN), which was confirmed by the loading tests in 2001.

Fig. 3.10 Weighing 6 tons by students and teachers.

3.8 EXERCISES

1. Describe the material properties of stone, wood, steel, and concrete, respectively.
2. For stone bridges, timber bridges, steel bridges, concrete bridges, list more than five famous bridges for each type, respectively.

REFERENCES

Troyano, L.F., 2003. Bridge Engineering: A Global Perspective, third ed. Thomas Telford, London.
University of Cambridge, 2016. The structure and mechanical behaviour of wood. Available from http://www.doitpoms.ac.uk/tlplib/wood/index.php.
Yoda, T., 1998/2004. Report on the field tests of Kintaikyo bridge. The Integrated Research Institute for Science and Engineering of Waseda University, Tokyo (in Japanese).

CHAPTER FOUR

Loads and Load Distribution

4.1 INTRODUCTION

Bridge structures are designed to carry traffic during their service lives. Bridge loads are actions in the form of forces, deformations, or accelerations applied to a structure or its components. The load acting on the bridge structures are generally divided into two categories: (1) those acting on the superstructure, and (2) those acting on the substructure. The major load components of highway bridges are dead load, live load (static and dynamic), environmental loads (temperature, wind, and earthquake), and other loads (collision, emergency braking).

The classifications of the load could be different according to the design specifications, but can be roughly divided into predominant (primary) load and subordinate (secondary) load. The load applied on bridge structures can also be classified as static load and dynamic load, as well as the concentrated load and distributed load etc.

Taking the Japan design load as an example, four load systems were divided according to the Standard Specification of Highway Bridges of Japan Road Association. They are:

1. *Principal loads (P)*—dead load (*D*), live load (*L*), impact load (*I*), prestressed forces (*PS*), concrete creep (*CR*), drying shrinkage (*SH*), earth pressure (*E*), hydraulic pressure (*HP*), and buoyancy or uplift (*U*).
2. *Subordinate loads (S)*—wind load (*W*), temperature change (*T*), and earthquakes (*EQ*).
3. *Special loads corresponding to principal loads (PP)*—snow load (*SW*), influence of ground displacement (*GD*), influence of support displacement (*SD*), wave pressure (*WP*), and centrifugal force (*CF*).
4. *Special loads corresponding to subordinate loads (PA)*—braking force (*BK*), erection load (*ER*), collision force (*CO*), others.

Bridge Engineering
http://dx.doi.org/10.1016/B978-0-12-804432-2.00004-9

According to the bridge location and bridge type, the above mentioned loads should be selected appropriately during the structural design and analysis, but not necessarily consider all the loads. Major loads considered in the bridge design are discussed below.

4.2 DEAD LOAD

Gravity loads of constant magnitudes and fixed positions act permanently on the structure. Such loads consist of the weights of the structural system itself and of all other material and equipment permanently attached to the structural system. In the bridge design, the dead load denotes the constant load in a bridge due to the weight of the members, the supported structure, and permanent attachments or accessories. To be specific, the dead load in a bridge include: (1) Facilities and additives (or accessories) on the bridge, such as guardrail, lamp standard etc.; (2) Self-weight of the deck system, such as deck, pavement, and pedestrian etc., (3) Self-weight of the floor system, such as stringer, transverse beam etc., and (4) Self-weight of the main girder or main structure system, including the floor beam etc. Although it is possible to determine (1) and (2) before the main structure design, it is hard to determine (3) and (4) before the final design of the main girder. Typical specific weights of different materials are summarized in Table 4.1.

4.3 LIVE LOAD

Live load in bridge design generally refers to loads due to moving vehicles that are dynamic, or the loads that change their positions with respect to time. This is unlike building structure, where live loads are the occupancy loads, which are considered as static load (Taly and Taly, 1997). The live load has been increasing with the progress of time. For modern bridges, their

Table 4.1 Specific Weight of Different Materials (kN/m^3) (JRA, 2012)

Material	Unit Weight	Material	Unit Weight
Steel	77	Concrete	23
Cast iron	71	Cement mortar	21
Aluminum	27.5	Wood	8.0
Reinforced concrete	24.5	Bituminous material (water proofing)	11
Prestressed concrete	24.5	Asphalt pavement	22.5

service lives are generally decades or even more than a hundred years. There-fore, the appropriate calculations or predictions for future service loads are necessary. Furthermore, the position of a live load may change, so each member of the structure must be designed for the position of the load that causes the maximum internal sectional forces inducing maximum stress in that member.

For highway bridges, the live load includes the vehicle load and sidewalk load. The live load generally varies according to bridge locations and the traffic conditions of the oversize vehicles. The design live load is diverse for different design specifications, and some representative live loads used in the engineering practice are listed below.

4.3.1 Live Load in US Specification

In the United States, the design vehicular live load is divided into three cat-egories, namely (1) design truck load, (2) design tandem load, and (3) design lane load.

The design truck specified in AASHTO Load and Resistance Factor Design (2007) are shown in Fig. 4.1. The weight of the front axle is 35,000 N with double rear axles weighing 145,000 N, respectively. The spacing between the two 145,000 N axles needs to be varied from 4300 to 9000 mm to produce the extreme force effects. The tire contact area of a wheel consisting of one or two tires is assumed to be a single rectangle with 510 mm in width and 250 mm in length.

The design tandem consists of a pair of 110,000 N axles spaced 1200 mm apart, with a transverse spacing of 1800 mm for the wheels. With regard to the design lane load, a uniformly distributed load of 9.3 N/mm shall be applied in the longitudinal direction with a width of 3000 mm in the trans-verse direction.

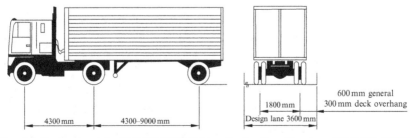

Fig. 4.1 Design truck in AASHTO LRFD Bridge Design Specifications (AASHTO, 2007).

In the design, the following loading conditions shall be considered to produce the extreme force effects: (1) the effects of the design tandem together with the design lane load, (2) the effects of one design truck with the variable axle spacing plus the design lane load, and (3) for negative bending moment regions, 90% of the effects of two design trucks with a minimum spacing of 15,000 mm between the lead axles of one truck and the rear axle of the other truck, plus the 90% of the effects of the design lane load. In this case, the spacing between the two 145,000 N axles is taken as 4300 mm.

4.3.2 Live Load in Japanese Specification

Both truck load and lane load used in Japan bridge design specification are shown in Figs. 4.2 and 4.3, respectively. The truck load, also denoted as T-load, is mainly used for designing the deck systems. The lane load, or L-load, is mainly used for designing the main girders.

For the lane load, the distributed load of p_1 and p_2 are taken based on the loading conditions of the oversized vehicles. A-Type live load is used for municipal roads with a small number of oversize vehicles, while the B-Type live load is used for national highway, a prefectural road (highway), and a main (principal, trunk) road with large number of oversize vehicles. In addition, the loading intensities of p_1 and p_2 also vary according to the internal forces including shear force or bending moment, as indicated in Table 4.2.

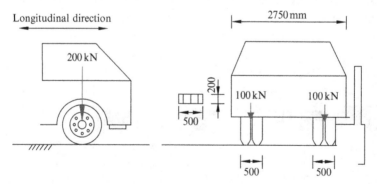

Fig. 4.2 T-load in Japanese bridge design specification (unit: mm).

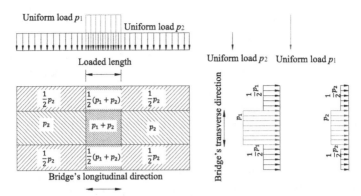

Fig. 4.3 L-load in Japanese bridge design specification.

Table 4.2 Distributed Load Intensities of the Lane Load

Live Load on Main Lanes (Width: 5.5 m)

Load	Load Length D (m)	Uniform Load p_1 (kN/m²)		Uniform Load p_2 (kN/m²)			Live Load on Secondary Lanes
		For Bending Moment	For Shear Force	L ≤ 80	80 < L ≤ 130	130 ≤ L	
A-Live Load	6	10	12	3.5	4.3–0.01 L	3.0	50% of live load on main lanes
B-Live Load	10						

4.3.3 Live Load in British Specification

In the British standard code for the design and construction of steel, concrete, and composite bridge (BS 5400-2, 2006), two types of loading are generally considered in the design of highway bridge including HA and HB types of loading. Both HA and HB loading includes the effect of impact for which the HA loading represents normal traffic effect and HB loading represents abnormal vehicle unit loading effect. The HA loading is the combination of uniform distributed load (UDL) with knife edge load (KEL), or a single nominal wheel load. The UDL has the magnitude depending on the loaded length specified in the standard and defined by the influence line analyses while the KEL has a magnitude of 120 kN per notional lane. The single nominal wheel load is specified as a 100 kN load placed on the carriageway and uniformly distributed over a circular contact area assuming an effective pressure of 1.1 N/mm². The HB loading is a series of four axle loads in which each axle consists of four wheel load 1 m apart. Both, two front axles and two back axles are 1.8 m apart with a set of variable distance between second and third axle to produce the most critical effect. As specified in the code, the minimum units to be considered are 25 and may go up to 45 if directed by the appropriate authority. The BS5400 was replaced by the Eurocodes for the newly designed bridges built after 2010 but still remains as the basis of the assessment standards for the existing bridges.

4.3.4 Live Load in European Specification

The Eurocode EN 1991-2(2003) specifies four loading models to be considered as the road traffic effects for the ultimate limit states with particular serviceability verifications. The Load Model1 (LM1) represents normal traffic which consists of Tandem System (TS) with double-axle concentrate loads 1.2 m apart (center to center) and UDL to include the effect of lorries and cars. The LM1 should be used for both general and local verifications. The TS and UDL have the intensity of $\alpha_Q Q_k$ (for each axle) and $\alpha_q q_k$, respectively. The adjustment factors α_Q and α_q are specified in the National Annex. The Q_k is distributed on two wheels 2 m apart (center to center) over the contact surface of 0.4 m × 0.4 m each. On each lane, only one TS (as shown in Fig. 4.4) should be used with at most three loading lanes. The intensity for each loading lane is specified in Table 4.3. The q_k should apply on all the unfavorite part of the influence surface, whose intensity is

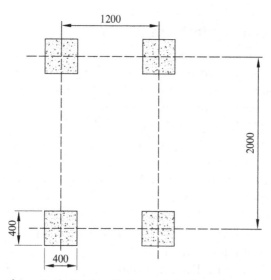

Fig. 4.4 Details of one single Tandem in Eurocode (unit: mm).

Table 4.3 Intensity of Axle Load for TS

Lane Number	Tandem System (TS) Axle Load Q_k (kN)
1	300
2	200
3	100
≥ 4	0

Table 4.4 Intensity of Uniform Distributed Load (UDL)

Lane Number (i)	Uniform Distributed Load (UDL) q_k (kN/m^2)
1	9
≥ 2 (and other remaining areas)	2.5

taken as $9\,\text{kN/m}^2$ for the defined primary lane and $2.5\,\text{kN/m}^2$ for the remaining area as specified in Table 4.4.

Load Model 2 (LM2) is a single axle load applied on specific tyre contact areas which covers the dynamic effects of the normal traffic on short structural member (EN 1991-2, 2003). The load has a magnitude of $\beta_Q Q_{ak}$ with $Q_{ak} = 400$ kN, and β_Q specified in the National Annex of each country adopting the Eurocode. Load Model 3 (LM3) is a set of assemblies of axle loads representing special vehicles (e.g., for industrial transport), which can travel on routes permitted for abnormal loads. It is intended for general and local verifications. Load Model 4 (LM4), generally known as crowd loading is represented by a Load model consisting of a uniformly distributed load (which includes dynamic simplification) equal to $5\,\text{kN/m}^2$.

4.3.5 Live Load in Chinese Specification

Both lane load and truck load are specified as live load in Chinese design specification, where the lane load is mainly used for designing the main girders, while the truck load is mainly used for checking the safety of the bridge deck.

The lane load used in Chinese bridge design specification is shown in Fig. 4.5, consisting of both concentrated load (P_k) and distributed load (q_k). In Chinese specification, two levels of highway including Grade A highway (or highway-I) and Grade B highway (or highway-II) are classified. In the design of Grade A highway, the distributed load (q_k) is taken as 10.5 kN/m, while the concentrated load (P_k) should be decided according to the length of the bridge superstructure. For a bridge with a span length smaller than 5 m or larger than 50 m, the concentrated load should be taken as 180 kN

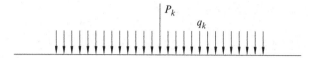

Fig. 4.5 Design lane load in Chinese specification.

and 360 kN, respectively. For a bridge span length from 5 to 50 m, the concentrated load is determined according to the first-order interpolation of the bridge length. For the design of Grade B highway bridge, the load is taken as 0.75 of the design load as specified for Grade A highway bridge.

The size dimension and transverse distribution of the design truck load specified in Chinese design specification are shown in Figs. 4.6 and 4.7, respectively.

4.4 IMPACT

When live loads are moving rapidly across a structure, they cause larger stresses than those that would be produced if the same loads would have been applied gradually due to the road roughness, expansion joint, and vibration of the engine etc. This dynamic effect of the load is referred to as impact in the bridge design. Live loads expected to cause such a dynamic effect on structures are increased by impact factors. The stress increase due to the impact is taken into account by using the following equation:

$$\sigma = \sigma_s + \sigma_d = \sigma_s(1 + i) \tag{4.1}$$

where i is the impact factor.

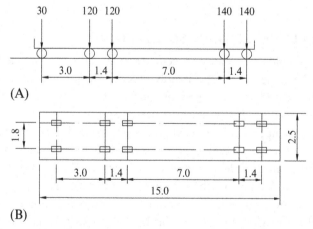

Fig. 4.6 Size dimensions of design truck in China (unit: m). (A) Front view. (B) Plan view.

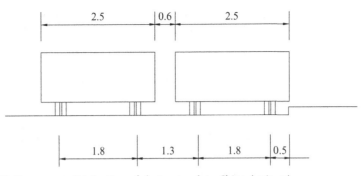

Fig. 4.7 Transverse distribution of design truck in China (unit: m).

The impact factor of a bridge is closely related to the bridge span length, structural type, the ratio between the dead load and the live load etc. In Japanese bridge design specification, the impact factor is determined according to the type of the bridges, as shown in Table 4.5. In addition, the span length should be taken accordingly based on the bridge type and structural type etc.

4.5 WIND

Wind loads are produced by the flow of wind around structures. In the bridge design, the wind load is defined as the wind pressure on the bridge. Wind load magnitudes vary with the peak wind speed, type of terrain etc. For large span bridges, especially the cable-stayed bridge and suspension bridge, wind load is an important design load and often play a critical role that affects the

Table 4.5 Impact Factor in Japan Bridge Design Specification

Bridge Type	Impact Factor	Note
Steel bridge	$i = \dfrac{20}{50 + L}$	Both truck load and lane load
RC bridge	$i = \dfrac{20}{50 + L}$	Truck load
	$i = \dfrac{7}{20 + L}$	Lane load
PC bridge	$i = \dfrac{20}{50 + L}$	Truck load
	$i = \dfrac{10}{25 + L}$	Lane load

strength, stiffness, and stability of the bridges. The significant role of wind loads is more highlighted after it caused damages to a number of bridge structures, some even collapsed completely, e.g., Tacoma Narrows Bridge (1940).

The design wind load for static design used in Japan is given by the following formula (JRA, 2012):

$$p = \frac{1}{2}\rho U_d^2 C_d G \qquad (4.2)$$

where ρ is the air density and generally taken as 1.23 kg/m^3, U_d denotes the design wind load (40 m/s), C_d stands for the drag coefficient, and G is the gust factor.

4.6 TEMPERATURE

The change in temperature will cause the deformation of the bridge. A determinate structure will expand or contract, but the strains generated by the change in temperature will not cause stress in its structural members. In an indeterminate structure, however, the stress caused by temperature changes may be comparable to that caused by live load due to the traffic (Catbas, 2008).

In general, two types of temperature including uniform change and gradient change occur in the bridge structure. The year round temperature change will result in uniform change, resulting in the bridge length change along the axis direction. When such deformation is constrained, the temperature induced forces (or thermal forces) will occur. On the other hand, the gradient change in temperature is mainly due to the solar radiation and the surrounding air. This will cause the nonlinear temperature change along the height direction, and results in stress on the section and further secondary forces if constrained.

4.7 SEISMIC LOAD

Seismic loading is one of the basic concepts of earthquake engineering which means application of a seismic oscillation to a structure. It happens at contact surfaces of a structure either with the ground or with adjacent structures. Seismic loading depends, primarily on seismic hazard, geotechnical parameters of the site, and structure's natural frequency etc. There are horizontal and vertical components of earthquake ground motions, but the horizontal component of earthquake ground motion is the main cause of bridge damages. Therefore, only horizontal earthquake ground motion is considered in the bridge design and analysis.

The basic seismic design philosophy for most of the codes is performance based design or capacity design (Buckle, 1996). In the Japanese design specification, bridges are divided into two important categories: ordinary bridges and important bridges. In the bridge design, the seismic performance of the bridge under corresponding seismic load should be carefully checked. The seismic design for highway bridges in Japan shall refer to the Part V (Seismic Design) of "Specifications for Highway Bridges."

4.8 SNOW AND ICE

In some places of the world, snow and ice are significant for considerable period and this should be considered in the bridge design. This is especially for large span bridges, such as cable-stayed bridges or suspension bridges, on which the snow is hard to be removed completely. In Japan, there are two cases in which SW should be considered: (1) vehicles can move freely on sufficiently compressed snow, or (2) vehicles cannot move freely due to the heavy snow coverage. For the first case, the SW with thickness of 150 mm is generally assumed and the load is taken as 1 kN/m^2. While, for the second case, the SW is determined according to the following equation:

$$SW = P \times Zs \qquad (4.3)$$

where SW is the snow load (kN/m^2), P is the mean weight of the snow (kN/m^3), and Zs denotes the design snow coverage depth (m). Although the weight of snow varies between regions and seasons, the design SW of 3.5 kN/m^2 may be generally used.

4.9 CONSTRUCTION LOAD

Temporary forces occur during the bridge construction stage due to the deadweight of the equipment or plant are called construction load. Construction load is dependent on the construction method and is different in each construction stage. Unconsidered construction load may cause the buckling or even collapse of the bridge superstructure, or severe damage of the substructure such as piers or foundations.

4.10 CREEP AND SHRINKAGE OF CONCRETE

Creep and shrinkage are two physical properties of concrete. For concrete bridge and steel-concrete composite bridge, the creep and SH of concrete shall be considered in the design.

Creep in concrete is the tendency of concrete material to deform under the influence of mechanical stresses. In ordinary bridge structures, the sustained load induced stress is generally less than 40% of the compressive strength of the concrete. If it is the case, the creep strain of the concrete can be taken by using the following equation:

$$\varepsilon_{cc} = \frac{\sigma_c}{E_c} \varphi \tag{4.4}$$

in which $\varepsilon_{cc}, \sigma_c, E_c,$ and φ are the creep strain, stress due to sustained load, Young's modulus, and creep coefficient, respectively.

There are two types of concrete shrinkage, including the shrinkage when the moisture in the concrete dissipates to the outside to dry and the shrinkage due to moisture consumption in the concrete by cement hydration (self-shrinkage). Both of these shrinkages should be considered in the bridge design.

4.11 COMBINATION OF LOADS FOR BRIDGE DESIGN

Besides the loadings listed above, other loadings including the prestress, braking force etc. should also be considered appropriately in the design and structural analyses. In addition, different load combinations should be considered in the bridge design. A load combination sums or envelopes the analysis results of certain load cases. Taking the Japanese bridge design specification as an example, the 10 load combinations and corresponding multiplier factors specified in Table 4.6 should be used to check the safety of the bridge structures (Nagai et al., 2014).

4.12 EXERCISES

1. Describe the "four-load systems" in the standard specifications of highway bridges in Japan.
2. What is the difference between "T" load and "L" load?
3. Describe the "earthquake ground motions" and "seismic performances requirement" used in the standard specifications of highway bridges.

Table 4.6 Design Load Combinations Associated With Multiplier Factors for Allowable Stress

No	Load Combination	Multiplier Factor
1	P + PP	1.00
2	P + PP + T	1.15
3	P + PP + W	1.25
4	P + PP + T + W	1.35
5	P + PP + BK	1.25
6	P + PP + CO	1.70 for steel members 1.50 for concrete members
7	$P' + EQ$	1.50
8	W	1.20
9	BK	1.20
10	ER	1.25

REFERENCES

AASHTO, 2007. AASHTO LRFD Bridge Design Specifications. AASHTO, Washington, DC.

BS 5400-2, 2006. Steel, concrete and composite bridges. Specification for loads.

Buckle, L.G., 1996. Overview of the seismic design methods for bridges in different countries and future directions. In: Eleventh World Engineering Conference on Earthquake Engineering, Paper Number: 2113.

Catbas, F.N., Susoy, M., Frangopol, 2008. Structural health monitoring and reliability estimation: long span truss bridge application with environmental monitoring data. Eng. Struct. 30 (9), 2347–2359.

EN 1991-2, 2003. Eurocode 1: Actions on structures – Part 2: Traffic loads on bridges. CEN (European Committee for Standardization), Brussels, Belgium.

Japan Road Association, 2012. Design Specification for Highway Bridges and Commentary, Part I: Common Part, Part II: Steel Bridges. Japan Road Association, Tokyo, Japan (Japanese).

Nagai, M., Okui, Y., Kawai, Y., Yamamoto, M., Saito, M., 2014. Bridge engineering in Japan. In: Handbook of International Bridge Engineering. CRC Press, Boca Raton, FL.

Taly, N., Taly, M., 1997. Design of Modern Highway Bridges. McGraw-Hill, New York, NY.

CHAPTER FIVE

Bridge Deck Systems

5.1 INTRODUCTION

A bridge deck (or road bed) is the roadway, or the pedestrian walkway, surface of a bridge. The deck may be of either cast-in-situ or precast concrete, wood which in turn may be covered with asphalt concrete or other pavement. The concrete deck may be an integral part of the bridge structure (e.g., T-section beam structure), or it may be supported with I-beams or steel girders, as so-called composite bridges. The deck may also be of other materials, such as wood or open steel grating.

Sometimes the deck system is called a floor system, such as for a bridge deck that installed in a through truss. A suspended bridge deck will be suspended from the main structural elements on a suspension or arch bridge. On some bridges, such as a tied arch or a cable stayed, the deck is a primary structural element, carrying tension or compression to support the span. But for girder beams, the bridge deck system is not the load carrying system. Despite this, they are important for the bridge serviceability, safety as well as the aesthetics. Thus, deck system deserves special attention in all bridge design and construction.

The deck system varies with different bridge types and bridge superstructure construction methods, and particular attention of this chapter will be given to the bridge accessories with special emphasis on pavement, drainage system and waterproofing system, expansion joint, sidewalk, lamps post, handrail, and guardrail.

5.2 LAYOUT OF THE DECK SURFACE

The layout of the bridge deck surface should be determined according to the deck width, the design speed, and the hierarchy of roads. In general, there are following three types.

5.2.1 Undivided Carriageway

Undivided carriageway denotes that the traffic load located at the same surface, also uplink and downlink, was not divided. As the motor vehicles and nonmotor vehicles on the same road surface, the traffic can only in middle or low speed, it can easily has traffic jam on the bridge.

5.2.2 Divided Carriageway

To avoid the possible traffic jam on the carriageway, the carriageway can be divided by using the median strip, or sometimes the uplink and downlink located at two bridges. The separation between the uplink and downlink, or different transportation means such as the motor traffic and nonmotor traffic makes it become easy to control the traffic and improve the traffic capacity.

5.2.3 Double-Decked Bridges

Double-decked bridges denote the bridges that have two levels deck system. Double decks were generally used for different means of transportation, which are useful for improving the traffic capacity and traffic control. In addition, such bridge can be used for reducing the bridge deck width and make full use of the clearance. Such as the Nanjing Yangtze River Bridge in Fig. 5.1, which is a double-decked road-rail truss bridge across the

Fig. 5.1 The Nanjing Yangtze River Bridge. *(Photo by Lin.)*

Yangtze River China. Its upper deck is part of China National Highway, and its lower deck carries a double-track railway.

5.3 BRIDGE PAVEMENT

The pavement is the important portion of the bridge deck that vehicles come in direct touch, and a structurally sound, smooth riding, and long lasting pavement is very important for bridge users. The bridge pavement is used for protecting the slab deck from the impact due to traffic load, rainwater, and other meteorological conditions, and providing durable and comfortable traffic conditions. A rough pavement is uncomfortable to the drivers, and a quality pavement should be designed and constructed according to appreciate design specifications for the pavement.

5.3.1 Functions and Requirement of Bridge Pavement

The main functions of bridge pavement include: (1) prevent the vehicle tie or the caterpillar track directly wear the bridge deck; (2) protect the bridge deck and main girder from water erosion; and (3) dispersion of the concentrated truck load.

Bridge pavement quality is important for the survivability and durability of the bridge structures. Nonetheless, bridge deck pavements must meet a large number of requirements related to strength, wear-resisting, crack-resisting, antiskid, and good integral with bridge deck. The bridge pavement shall have adequate resistance to permanent deformation, vehicle sliding without cracking, etc. It also must protect and seal the underlying supporting structure as this determines to the durability of bridge superstructure. The pavement should also be able to absorb traffic loads and transfer them to the deck and supporting structures but remain even within allowable deformation and provide good antiskid conditions for vehicles. Besides, they must protect the bridge structure from surface water.

5.3.2 Classifications of Bridge Pavement

Cement concrete pavement and asphalt pavement are most often used pavement method in bridge structures. Cement concrete pavement has advantages like less expensive, wear-resisting, suitable for high-traffic bridge, environmental sustainability, durability, and requires less repair and maintenance over time, but requires longer curing time. On the other hand, the asphalt pavement requires less curing time and has lighter weight, easy repair or replacement, but easy to get aging and deform.

A cement concrete pavement shall be constructed simultaneously with the slab concrete in order to form an integral structure. If casted separately with slab concrete, the pavement concrete will be vulnerable to drying shrinkage crack because of their relatively thin thickness. There is a concern of stripping due to bridge vibration, impact of vehicles, and rainwater permeation. For this reason, the pavement concrete and deck concrete should be constructed at the same time but not separately. If rainwater permeates the concrete slab, not only causing the corrosion of the reinforcements or any other structural steel in the concrete, but also accelerate the concrete deterioration, particular for the deck under repeated load in the service condition. These have remarkable effect on the durability and load carrying capacity of the bridge. Because of this, adequate sealing measures should be taken not only for the concrete deck but also for the members near to expansion joint and other accessory devices (Japan Road Association, 2012a,b).

In general, the asphalt bridge pavement system consists four different layers: a sealing layer, a waterproofing layer, a protecting layer, and surface layer (asphalt). For an asphalt pavement, a waterproofing layer should be used to present rainwater permeation. Although different application techniques and materials can be used on steel and concrete bridge decks, the general construction steps on a bridge deck starts by surfacing of the deck, followed by sealing layer, a waterproofing layer, a protecting layer, and the surface layer on top. The surface and subsurface drainage system should be applied on both steel and concrete decks. The sealing layer can be made from various materials, including bituminous materials (EAPA, 2013). The bridge pavement surface is generally built as a parabola curve with the cross slope of 1.5%–2%. The sidewalk pavement surface is usually built as straight line with a cross slope of 1%.

5.4 DRAINAGE SYSTEM

In addition to installing the waterproof layer in the pavement, the water on the bridge must be drained quickly from the bridge pavement to ensure the vehicular traffic safety. The bridge deck drainage system includes the bridge deck itself, inlets, bridge gutters, and drainage pipes.

If superelevations and cross-slopes of bridge surface are properly designed, water and dust can be efficiently removed away. In general, the transverse surface of bridge deck and gutters are designed as 1.5%–2.0% as standard. Bridge deck designs with zero grades may cause water erosion problems. Then, additional facilities like catch basins should be installed

Fig. 5.2 Drainage pipes in bridges. *(Photos by Lin.)*

at proper intervals (e.g., 20 m) considering longitudinal and transverse slopes. A catch basin should also be used near the expansion joint to minimize the inflow (Japan Road Association, 2012a,b).

Drainpipes (Fig. 5.2) are generally designed as circular shape with a minimum diameter of 150 mm, which will be good to avoid bends. Drain pipes generally extend the length of 15–20 cm from the bridge bottom, and the wind direction should be considered to avoid blowing the water into the main girder. In addition, as drainpipes are less durable than the bridge itself, drainpipes should be designed with consideration of possible inspection, maintenance, or replacement in service stage.

5.5 WATERPROOFING SYSTEM

Bridge decks are vulnerable, subject to attack by water and chloride that can lead to deterioration and issues with longevity and durability of the bridge decks. The installation of an effective waterproofing membrane is therefore an essential part of bridge deck system in addition to the drainage system. It represents the protecting wall against the water and aggressive chemicals that would corrode the steel reinforcing bar and the concrete in concrete decks. For steel decks, the asphalt pavement on the top of the main girder or the longitudinal stiffener is vulnerable to cracks, where the waterproofing is also indispensable. In addition to the deck, the waterproofing should also be used for any interconnection parts of the bridge deck, such as an expansion joint.

For bridge decks, several waterproofing systems are available, such as sheet systems, liquid systems, and mastic asphalt system. However, the mastic asphalt system has been used in the past but is rarely used recently.

5.6 BRIDGE EXPANSION JOINT

5.6.1 Functions of Expansion Joint

Bridge expansion joints are designed to adjust its length accommodating movement or deformation by external loads, shrinkage, or temperature variations, and allow for continuous traffic between bridge structures and interconnecting structures (another bridge or abutment). The expansion joints can also be used for reducing internal forces in extreme conditions and allow enough vertical movement for bearing replacement. Steel expansion joints are most commonly used, though rubber joints are also often used to provide a smooth transition for modern bridge construction, or continuous girders (Toma et al., 2005).

It was suggested that expansion joints fall into three broad categories depending upon the amount of movement accommodated (Malla and Shaw, 2003), including: (1) small movement joints capable of accommodating movement up to about 45 mm; (2) medium movement joints capable of accommodating total motion ranges between about 45 mm and about 130 mm; and (3) large movements joints include systems accommodating total motion ranges in excess of about 130 mm. There are many different types of expansion joints according to JASBC (1984), such as blind type, slit plate type, angle joint type, postfitting butt type, rubber joint type, steel-covered plate type, and steel finger type. According to ICE (2008), there are buried joints, asphaltic plug joints, nosing joints, reinforced elastomeric joints, elastomeric in metal runners joints, and cantilever comb or tooth joints. Some examples of expansion joints used in bridge structures are shown in Fig. 5.3.

Expansion joints should be installed as late as possible in bridge constructions allow for shrinkage, creep, and settlement movements to have taken

Fig. 5.3 Different bridge expansion joints. *(Photos by Lin.)*

place. Similar to other semipermanent members, the expansion joints should be designed so as to be easily replaced or reset in service stage.

5.6.2 Dynamic Behavior of Bridge Expansion Joints

If a bridge is subjected to a dynamic moving load, the response varies with the time. Until recently, the dynamic requirements of bridge expansion joints to be taken into consideration were relatively less important. The design methods of the expansion joints were meant to deal with the structural problem in statistical terms by using the so-called dynamic factor or impact factor. The role of expansion joints is to carry loads and provide safety to the traffic over the gap between a bridge and an abutment or between two bridges. A further requirement is a low noise level especially in densely populated area. Therefore, expansion joints should be robust and suitable for static and dynamic actions. Movements of expansion joints depend largely on the size of the bridges and the arrangement of the bearings.

In design phases, expansion joints are required to have movement capacity, bearing capacity for static and dynamic loading, water-tightness, low noise emission, and traffic safety. On the basis of the fact that the failure due to impact loading is the main reason for the observed damages, we shall focus our attention to the impact factor for vehicle load that is governed by traffic impact because it differs from the static loading. The cantilever-toothed aluminum joint (finger joint) is one of the promising joints under impact loading to overcome this difficulty.

From the viewpoint of design methodology, numerical studies for impact behavior were conducted for aluminum alloy expansion joints with perforated dowels. The design impact factor for the expansion joints with the perforated dowels against traffic impact loading was examined by using numerical simulations, in which the dynamic amplification factor defined as the ratio of dynamic to static response was compared at various input-load patterns to the factors for expansion joints (Figs. 5.4 and 5.5).

The mechanical characteristics can be considered as indicators of the dynamic behavior so that the durability of an expansion joint is nothing more than the maintenance of performance in time. The value of frequencies and damping of the different modes is the main indicator. In laboratories, these characteristics can be more or less simulated, but the correlation with the situ behavior is not so easy. More precise methods of impact effect control must be established to facilitate the appreciation of the stage of the cumulative damage.

Fig. 5.4 Bridge expansion joint.

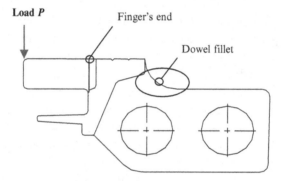

Fig. 5.5 Aluminum alloy expansion joints.

5.7 UNSEATING PREVENTION SYSTEM

The Bridge Collapse Prevention Device connects bridge superstructure with bridge piers, abutment, or adjacent girders by PC cables to prevent
bridge collapse in case of larger displacement during an earthquake. Following the 1995 Kobe Earthquake, the Japanese specifications for highway
bridges were revised a lot, and bridge-fall prevention devices have been
emphasized as important earthquake-resistant measures for bridge structures
in Japan. Bridge-fall prevention devices are used as a part of seismic-resistant
systems together with bearings and expansion joints.

An unseating prevention system consists the seating length, unseating prevention structures, and constructions for preventing excessive displacement

in the transverse direction to the bridge axis. An unseating prevention system must be selected according to the bridge type, bearing type, and ground conditions.

5.7.1 Seating Length

Seating length of a girder at its support shall be determined according to the following equations. The seating length of a girder at its support shall not be less than the value obtained from Eq. (5.1). In the case where the length is shorter than that obtained from Eq. (5.2), the design seating length shall not be less than the value from the latter equation. When the direction of soil pressure acting on the substructure differs from the longitudinal direction to the bridge axis, as in cases of a skew bridge or a curved bridge, the seat length shall be measured in the direction perpendicular to the front line of the bearing support.

$$S_{ER} = u_R + u_G \; (\text{m}) \tag{5.1}$$

$$S_{EM} = 0.7 + 0.005l \; (\text{m}) \tag{5.2}$$

$$u_G = \varepsilon_G L \; (\text{m}) \tag{5.3}$$

where S_{ER} is the required seating length (m), u_G is the relative displacement of the ground caused by seismic ground strain (m), u_R is the maximum relative displacement between the superstructure and the substructure crown due to Level 2 Earthquake Ground Motion (m), u_G is the relative displacement of the ground caused by seismic ground strain (m), S_{EM} is the minimum seating length (m), ε_G is the seismic ground strain, which can be assumed as 0.0025, 0.00375, and 0.005 for ground types I, II, and III, respectively, L is the distance between two substructures for determining the seating length (m), and l is the span length (m).

The seating length of a bridge should always no smaller than the required value, as shown in Fig. 5.6. Some bridge fall preventing devices are shown as examples in Fig. 5.7.

5.7.2 Unseating Prevention Structure

Ultimate strength of an unseating prevention structure shall not be less than the design seismic force determined by Eq. (5.4). The design allowance length of the unseating prevention structure shall be taken as large as possible with a maximum value determined by Eq. (5.5).

$$H_F = 1.5R_d \tag{5.4}$$

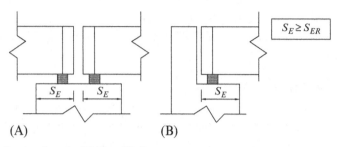

$$S_E \geq S_{ER}$$

(A)   (B)

Fig. 5.6 Seating length. (A) Pier. (B) Abutment.

Fig. 5.7 Bridge fall preventing device. *(Photos by Lin.)*

$$S_F = c_F S_E \; (\text{m}) \qquad\qquad (5.5)$$

where H_F is the design seismic force of the unseating prevention structure (kN), R_d is the dead load reaction (kN), S_F is the maximum design allowance length of unseating prevention structure (m), and c_F is the design displacement coefficient of unseating prevention structure, with the standard value of 0.75.

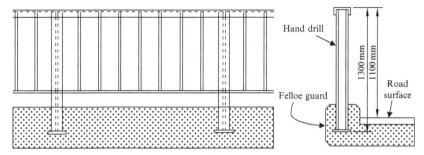

Fig. 5.8 Guard railings (Tachibana, 2000).

In addition to the above, constructions for preventing excessive displacement in the transverse direction should be appropriately considered.

5.8 GUARD RAILINGS

Guard railing, also called guard rail, handrail, or guardrail, is a system designed to keep people or vehicles from falling off the bridge. They may be a handrail for pedestrians, a heavier guard for vehicles, or a common railing for both. In general, the railings can be made of concrete, steel, or aluminum. The component slightly above the deck called felloe guard is used for fixing the hand rail, as shown in Fig. 5.8. As for its height, it is generally taken as 110 cm from the road surface. A thrust force of 2.5 kN/m is considered in horizontal direction for designing guard rails (Tachibana, 2000).

The guard railings are located prominently and are thus open to the critical eye of the public. It is important that they not only keep traffic within boundaries but also add to the aesthetic appeal of the whole bridge (Toma et al., 2005).

5.9 OTHER ACCESSORIES

There are many other accessories are necessary for overall function of a bridge, such as lamp post (Fig. 5.9), noise barriers, emergency telephone, nameplate, and they shall be installed accordingly as required. However, the construction of those accessories shall affect little on the main bridge structure and shall follow appropriate design codes.

Fig. 5.9 Lamp post examples. *(Photos by Lin.)*

 5.10 EXERCISES

1. Describe the components of the bridge deck system.
2. Describe the functions and requirement of the bridge pavement.
3. Describe the components of the bridge deck drainage system.
4. Why the expansion joints are necessary for the bridge structures?

REFERENCES

European Asphalt Pavement Association (EAPA), 2013. Asphalt pavements on bridge decks. EPEA position paper.

Institution of Civil Engineers, 2008. ICE Manual of Bridge Engineering, second ed. Thomas Telford Ltd.

Japan Road Association, 2012a. Design Specification for Highway Bridges and Commentary, Part I: Common Part, Part II: Steel Bridges. Japan Road Association, Tokyo (in Japanese).

Japan Road Association, 2012b. Design Specification for Highway Bridges and Commentary, Part I: Common Part, Part V: Seismic Design. Japan Road Association, Tokyo (in Japanese).

JASBC, 1984. A Guide Book of Expansion Joint Design for Steel Bridges. Japan Association of Steel Bridge Construction, Tokyo.

Malla, R.B., Shaw, M., 2003. Sealing of Small Movement Bridge Expansion Joints (Project Number: NETC 02-6). University of Connecticut.

Tachibana, Y., 2000. In: Nakai, H., Kitada, T. (Eds.), Bridge Engineering. Kyoritsu Shuppan Co., Ltd, Tokyo, Japan.

Toma, S., Duan, L., Chen, W.F., 2005. Chapter 25: bridge structures. In: Chen, W.F., Lui, E.M. (Eds.), Handbook of Structural Engineering, second ed. CRC Press, Boca Raton, FL.

CHAPTER SIX

Reinforced and Prestressed Concrete Bridges

6.1 INTRODUCTION

Concrete is a mixture of sand, gravel, crushed rock, or other aggregates held together in a rocklike mass with a paste of cement and water (McCormac and Brown, 2009). In practice, some admixtures can be added to change its hardening time (e.g., rapid hardening concrete), workability, durability etc. Like stone, concrete has a relatively high compressive strength but a very low tensile strength. To improve the performance of concrete, reinforced concrete and prestressed concrete were created and widely used in bridge construction. Reinforced concrete is a composite material, with the combination of concrete and steel reinforcement wherein the reinforcement provides the tensile strength lacking in the plain concrete. Steel reinforcements are also capable of resisting compression forces, and are used in columns as well as in other structures. For longer span concrete bridges, prestressing tendons provide a compressive load which produces a compressive stress to balance the tensile stress of concrete members due to a bending moment or other forces. Traditionally, the reinforced concrete is accomplished by using steel reinforcement inside poured plain concrete. The prestressed concrete can be classified into either pretensioned or posttensioned concrete (bonded or unbonded).

The first known use of concrete is for a floor in Israel, dating back to approximately 7000 BC (Domone and Illston, 2010). In the 2nd century BC, the Romans discovered that adding pozzolana to the lime produced a much stronger concrete, which could be used as a building material in its own right. This discovery allowed them to revolutionize construction by designing large span concrete domes, such as Hagia Sophia built in the 6th century, which is still in good condition, having resisted the ravages of the Mediterranean weather or even the earthquake (Benaim, 2007). However, the real breakthrough for concrete occurred in 1824, when an

English bricklayer obtained a patent for a cement that he called Portland cement (McCormac and Brown, 2009).

Concrete is one of the most frequently used construction materials for modern bridges all over the world due to the benefits of low cost, durability, aesthetics, and rapid construction techniques etc. First of all, the design of concrete bridges comes in all shapes and sizes, can almost meet different functional, aesthetic, and economic criteria, making concrete a construction material suitable for the most challenging structures. For instance, concrete bridges can be built in slabs, beams, box-girders, arches, cable-stayed suspension bridges with different span lengths. Secondly, concrete is an ideal material suited for use in aggressive environmental conditions, and require very little maintenance in the service condition, therefore providing a sustainable solution for bridge construction. Thirdly, the life cycle cost of concrete bridges is competitive due to the low initial construction cost and maintenance cost. Concrete is one of the cheapest construction materials, requires little inspection and maintenance, and demonstrates attractive cost efficiency in both construction and in service.

6.2 MATERIALS

The materials used in concrete bridges include the concrete, reinforcement, and prestressing bars etc., and their material properties are discussed below.

6.2.1 Concrete

The 28 days compressive strength of concrete is usually obtained from cylinders with a height to diameter ratio of 2 (such as a standard 150 mm diameter by 300 mm high cylinder) loaded longitudinally to failure. In addition to the cylinder specimen, the cube specimens are also used for concrete test. Fig. 6.1 presents typical stress–strain curves from unconfined concrete cylinders under uniaxial compression loading. The strain corresponding to the peak stress is about 0.002 and maximum compressive strain is approximately 0.003. The elasticity modulus of concrete is generally suggested in the specifications, or can be calculated as:

$$E_c = 0.043\gamma_c^{1.5}\sqrt{f_c'}\text{MPa} \tag{6.1}$$

where γ_c is the density of concrete (kg/m^3), and f_c' is the concrete strength (MPa).

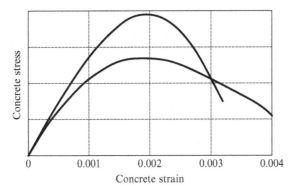

Fig. 6.1 Stress–strain curves of concrete under uniaxial compression loading.

The tensile strength of concrete can be determined directly from tension tests or indirectly in terms of the computed tensile strength. The elasticity modulus of concrete may be assumed to be the same for both tension and compression. Tension test for concrete is not often performed due to the different specimen holding patterns and the secondary stresses induced from the holding devices. Instead, the direct tensile strength of the normal concrete may be estimated as 10% of its compressive strength. In bridge design practice, however, the concrete tensile strength is generally ignored for safety considerations.

Creep and shrinkage are also important properties for concrete. Creep of concrete is referred to the behavior that the concrete deformation increases with time when it is subjected to a constant load. Creep of concrete is also an important reason for prestress loss in the prestressing tendons. Shrinkage of concrete is defined as the contraction due to loss of moisture caused by loss of water from voids or the reduction of volume during carbonation. Like creep, shrinkage is also important factor in prestress loss.

6.2.2 Reinforcement

Deformed steel bars with surface rolled with lugs or protrusions (for restricting longitudinal movement) are commonly employed as reinforcement in reinforced concrete bridge construction. Typical stress–strain curves for steel reinforcing bars may be obtained from the tension tests. The stress–strain curves exhibit an initial linear elastic portion, a yield phenomenon (stress increases little with the increase of strain), and a strain hardening (stress again increases as the strain increases). The Young's modulus of steel reinforcement E is generally taken as 200,000 MPa. The yield strength (stress at the yield point) is one of the most important data obtained from the tensile test and referred to in the bridge design. In a typical stress–strain curve for reinforcement (the same for structural steel), there are upper yield strength and lower yield strength. As the upper yield strength can be affected by several uncertainties such as loading speed, cross-section etc., the lower yield strength is often used in the structural design. Generally, the stress–strain curves for steel (both reinforcement and structural steel) are assumed to be identical in both tension and compression.

6.3 REINFORCED CONCRETE BRIDGES

Although concrete bridges can be built in different types such as arch bridges, cable-stayed bridges, and suspension bridges etc., they are more often used in medium span or short span beam bridges. Depending on the span length, the bridges may be built as reinforced concrete or prestressed concrete. The use of prestress provides effective connection and assembly method, making several rapid construction methods (e.g., balanced cantilever and incrementally launched method) became possible. The concrete beam bridges can be classified according to the cross section, the supporting condition, or construction method etc.

6.3.1 Cross Section

According to the cross section, the concrete beam bridges can be classified into slab bridge, T-beam bridge, and box-girder bridge, as shown in Figs. 6.2–6.4.

6.3.1.1 Slab Bridge

The slab bridges have the simplest superstructure configuration, and are very simple in design and construction, and less expensive. However, they are generally less efficient in use of construction materials, and only suitable for short span bridges. In addition to the solid section, the slab bridge can also be built as hollow section (e.g., voided section or cellular section shown in Fig. 6.2B) to reduce the deadweight and improve the material efficiency. Simply reinforced concrete hollow slab bridge with a span of 6–13 m, and prestressed concrete hollow slab bridge with a span of 10–20 m, have been found to be economical.

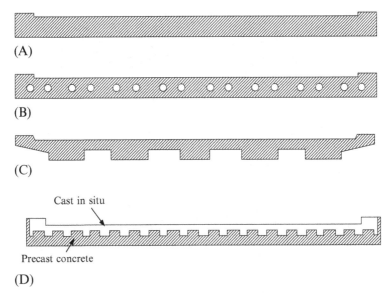

Fig. 6.2 Concrete slab bridge.

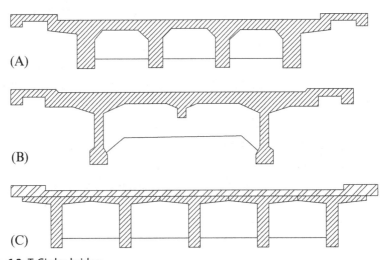

Fig. 6.3 T-Girder bridge.

6.3.1.2 T-Girder Bridge

The T-section is also named as ribbed slab, and it can be treated as a special design of hollow slab section; the concrete in tension is inefficient to sustain the load, and can be cut to reduce the deadweight. But in order to provide section for placing the reinforcing bars and prestressing reinforcement,

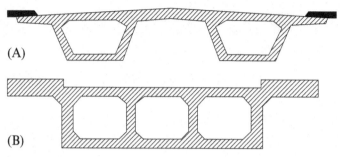

Fig. 6.4 Box-girder bridge.

certain area of cross section is still necessary. T girder bridges are generally very economical for spans of 12–18 m with girder stem thickness usually varying from 35 to 55 cm controlled by the necessary spacing of reinforcements.

6.3.1.3 Box-Girder Bridge

Box-girder sections consist of top deck, vertical web, and bottom slab. As both top and bottom flanges can resist stress, the box-girder can bear both positive and negative bending moments. The web can be relatively thin to reduce the deadweight. The box section is widely used for continuous, cantilever, and cable-stayed bridges. It has relatively large bending rigidity, torsion rigidity, and better load redistribution under eccentric load. Reinforced box-girder bridges are suitable for spans from 15 to 36 m.

6.3.2 Supporting Conditions

6.3.2.1 Simply Supported Bridges

The concrete beam bridges can be classified into simply supported, continuous, or cantilever bridges, as shown in Fig. 6.5. The simply supported bridges are statically determinate structures, they are very simple in design and they are widely used in short span bridges.

6.3.2.2 Continuous Bridges

The continuous spans are statically indeterminate structure, and the sections at intermediate support need to resist relatively large positive bending moment, resulting in the reduction of the positive bending moment in the span center section. The continuous span bridge has relatively large span capacity and is suitable for good foundations.

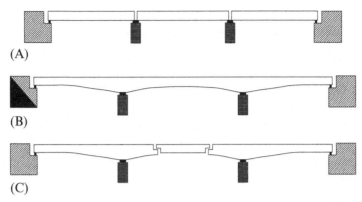

Fig. 6.5 Concrete bridge classifications according to supporting conditions. (A) Simply supported bridge. (B) Continuous bridge. (C) Cantilever bridges.

6.3.2.3 Cantilever Bridges
A cantilever bridge is a statically determinate structure, has larger span capacity than that of a simply supported bridge, but less span capacity than a continuous bridge.

6.3.3 Construction Method
Based on the construction method, the concrete beam bridges can be classified into precast bridge, case-in-place bridges, or a combination of the two methods.

6.3.3.1 Case-in-Place
The case-in-place bridges are constructed fully in its final location, thus have relatively good structural integrity, and can be constructed into different shapes. However, this method generally requires a large amount of false work and formworks.

6.3.3.2 Precast Concrete Bridge
The concrete bridges can also be precast, which means the bridges were built at other locations and then transported to construction site for placement in the whole bridge structure. This construction method has several advantages. Firstly, this method can have higher construction speed because the piling and member fabrication can be performed simultaneously, more efficient production in suitable and well organized site (factory), and is less affected or not affected from season to season. Also, this method has better quality control because of the established procedures by

"factory production. However, as there is a distance from fabrication yard to bridge location, transportation could be a problem. The transportation facilities and lifting equipment, element weight/size must be considered.

6.3.3.3 Combination of Cast-in-Place and Precast

For the combination type, some load carrying members are prefabricated, while the other members are casted in place. The prefabricated members are used as the false work for the cast–in–place members, and the cast–in–place members connect the different prefabricated members. The bridge structural integrity can be enhanced and all the members can carry the service load together.

6.4 PRESTRESSED CONCRETE BRIDGES

Prestressed concrete is a form of reinforced concrete that builds in compressive stresses during construction to oppose those found when in use. In other words, it is a combination of steel and concrete that takes advantage of the strengths of each material.

6.4.1 Prestressing Systems

The prestressing system is composed of a wire, strand and tendons, anchorage, stressing jacks, corrosion protection, and ducting etc., and images of the prestressing are shown in Figs. 6.6 and 6.7. The wire diameters are typically between 5 and 7 mm with a minimum tensile strength of 1570 N/mm^2 and carry forces up to 45 kN. Strand is an assembly of several high strength steel wires wound together. Anchorages are devices used at each end of the tendon, and forces are transferred into the concrete by an anchorage system after being tensioned by stressing jacks.

Fig. 6.6 Prestressing systems (anchorage, strand, and stressing jacks, etc.).

Fig. 6.7 T-section girder with prestressing. *(Photo by He.)*

Corrosion protection and ducting enable protecting the tendons from corrosion by coating a soluble oil etc. Posttensioned tendons are normally placed inside ducting systems made of aluminum or plastic etc.

6.4.2 Pretensioned and Posttensioned Concrete

There are two methods of prestressing in concrete bridges, including pretensioned prestressing and posttensioned prestressing.

6.4.2.1 Pretensioned Concrete

Pretensioning involves tensioning the cables before the concrete is poured in the formwork. After the concrete hardening, the tendons are cut (or released), resulting in the transmission of the prestress from tendons to concrete.

Pretension is the easiest control of the bonded stressing with the least chance of error in the bonding process. Tension caused by the steel is spread throughout the length of the concrete since it is bonded within the concrete along the length of the member. This method can produce a good connection between the concrete and tendon, which is important for protecting the tendon from corrosion and for allowing direct transmission of tension forces. However, as it requires strong anchoring points to stretch the tendon, pretensioned concrete members are generally prefabricated in the factory and then transported to the construction site. This is good for concrete curing, but may limit the size of concrete members because of the transportation capability. This technique is used principally for the construction of concrete bridges with short spans or structural members.

6.4.2.2 Posttensioned Concrete

In posttensioned construction, the tendons are stretched after the concrete is casted. Plastics or metal tubes or conduits with unstressed tendons are fixed in the formwork and then concrete is placed. After the concrete gets sufficiently hardened, the tendons are stressed by jacks and mechanically attached to anchorage devices to keep the tendons in their designed positions.

In posttensioned construction, tendons can either be bonded or unbonded to the surrounding concrete. The tension in the tendons is transferred continuously along its length through the surrounding concrete with a cementitious matrix, which surround the strand and commonly referred to as grout. It acts with the duct, which is encased in the concrete member to complete the bond path between the prestressing strands and the concrete member. After stressing of the tendon, the grout is injected into the void of the tendon duct, which houses the prestressing strands. Unbonded tendon is that by design, it does not form a bond along its length with the concrete. Unbonded tendons are generally covered with corrosion inhibiting coating and encased in a plastic sheathing. The force in the stressed tendon is transferred to the concrete primarily by the anchors provided at its ends (Aalami, 1994).

6.4.3 External Prestressing

External prestressing is a kind of postprestressing method, in which prestressing tendons are located outside the concrete section and the prestressing force is transferred to a structural member through end anchorages or deviators, as shown in Fig. 6.8. There are many advantages of this method, such as easy in concrete casting, thus better concrete quality, allows easy monitoring and replacement of tendons, and little requirement of concrete web thickness. External prestressing method is not only widely used in the construction of new bridges, but also in strengthening, repair, and rehabilitation of existing structures.

6.4.4 Losses of Prestress

The most important factor in prestressed concrete is the prestressing force; while the reduction of initial applied prestress to an effective value is the prestress loss. Prestress loss is an important problem because it will affect service ability and ultimate load carrying capacity of the prestressed concrete bridges. In general, prestress loss can be divided into two categories; namely, the

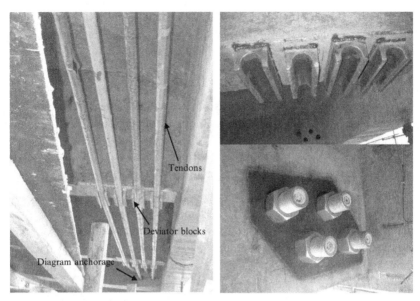

Fig. 6.8 Application of external prestressing in highway bridges. *(Photo by Ding.)*

short–term (or referred to as immediate) losses or long–term (or time dependent) losses.

Short–term prestress loss is also referred to as immediate loss, which represents the immediate losses that occur during prestressing of tendons. Short–term prestress loss is mainly due to the elastic shortening of concrete, friction between tendon and tendon duct, slip at anchorages after prestressing, and wobble effect etc.

Long–term prestress loss is also called time dependent prestress loss, which occurs during service stage of a bridge. Long–term prestress loss is mainly caused by creep and shrinkage of concrete, and relaxation of prestressing steel etc.

It is worth noting that the reasons for prestressing loss depend on the prestressing methods, like the prestress loss due to friction is only considered in posttensioned but not pretensioned concrete. For posttensioned, the prestressing loss due to elastic shortening is only taken into account when tendons are tensioned separately.

6.4.5 Construction Method

Similar to reinforced concrete bridges, the prestressed concrete bridges can be divided into cast in situ and precast depending on the construction location. Prestressed concrete bridges were more and more built as segmental bridges, which mean the bridge is built in short sections or segments, either cast in situ or precast.

6.4.5.1 Cast In Situ

Methods that are often used for cast in situ include the cast in situ posttensioned method, the balanced cantilever, and incrementally launched methods etc.

Posttensioned method is a cast in situ method which applies stressing after concrete hardening. This method is easy to operate but a large number of falsework and formworks are necessary. A highway concrete bridge built in this method is shown in Fig. 6.9. Balanced cantilever construction is an economical method when cast in situ with formworks is expensive or access from bridge below is practically impossible. Construction starts from the top of a pier, and the concrete segment is normally fixed to the pier either permanently or temporarily during the construction. This method can be used for both cast in situ and precast concrete bridge constructions. After a segment is cast, the formwork moves for the construction of the next segment. The segments on both sides can be built simultaneously so that the unbalanced moment is kept to a minimum. Also, rapid construction becomes possible by doing this. A temporary support may be used to improve the stability of the bridge during construction. A concrete bridge constructed by using balanced cantilever construction method is shown in Fig. 6.10.

In incrementally launched method, the concrete is cast in segments behind the abutment, the deck is pushed or pulled out over the piers. A specially prepared casting area is located behind the abutment with sections to assemble the reinforcement, to concrete, and to launch. As the deck is launched over a pier, large cantilever moments occur until the next pier is

Fig. 6.9 Posttensioned concrete bridge, cast-in-place. *(Photos by Ding.)*

Fig. 6.10 Balanced cantilever construction. *(Photo by Gui.)*

reached and to reduce these moments a temporary lightweight steel launching nose is fixed to the front of the girder (typically 60% of the span length and with the stiffness of about 10%–15% of the concrete deck). This method is generally used for spans of up to 60 m, the technique has been used for longer spans up to 100 m with the help of temporary piers placed to reduce the effective span length during launch.

6.4.5.2 Precast

Prestressed concrete bridges can also be precast in the factory and then moved to the construction site. In the last few decades, the precast concrete segmental bridge construction has been widely used around the world. These construction methods can benefit by reduction of costs, construction time, environmental impacts, and the maintenance of traffic. They also offer additional structural advantages of durability, fire resistance, deflection control, better rider serviceability, insensitivity to fatigue, and other redundancies (Kristensen, 2002). Precast segmental erection techniques for concrete bridges include the erection on falsework, erection by gantry, erection by crane, erection by lifting frame, and full span erection techniques etc. A precast concrete bridge with T-beams under construction is shown in Fig. 6.11.

Fig. 6.11 Precast T-beams. *(Photo by He.)*

6.5 EXERCISES

1. Describe the classification of concrete bridges according to with and without prestressing, construction methods, cross sections, and span types.
2. Describe the prestressing system used in bridge structures.
3. Describe the main types of prestressed-concrete structures.
4. Describe the primary and secondary effects of prestress, respectively.
5. Describe the types and reasons for prestress loss.

REFERENCES
Aalami, B.O., 1994. Unbonded and bonded post-tensioning systems in building construction, a design and performance review. Post-Tensioning Institute, Phoenix, AZ Technical Note, Issue 5.
Benaim, R., 2007. The Design of Prestressed Concrete Bridges: Concepts and Principles. CRC Press.
Domone, P., Illston, J., 2010. Construction Materials, fourth ed. Spon Press.
Kristensen, J.E., 2002. Precast Segmental Bridge Construction Part 1—An Introduction, A SunCam online continuing education course.
McCormac, J.C., Brown, R.H., 2009. Design of Reinforced Concrete, eighth ed. John Wiley & Sons.

CHAPTER SEVEN

Steel Bridges

7.1 INTRODUCTION

Steel bridges are widely used around the world in different structural forms with different span length, such as highway bridges, railway bridges, and footbridges. The main advantages of structural steel over other construction materials are its strength, ductility, easy fabrication, and rapid construction. It has a much higher strength in both tension and compression than concrete, and relatively good strength to cost ratio and stiffness to weight ratio. Steel is a versatile and effective material that provides efficient and sustainable solutions for bridge construction, particularly for long span bridges or bridges requiring enhanced seismic performance.

The structural steel for steel bridges should be selected according to the required material properties or the stress state where used, environmental conditions at the construction site, corrosion protection method, construction method, etc. (JSCE, 2007). The physical properties of structural steel such as strength, ductility, toughness, weldability, weather resistance, chemical composition, shape, size, and surface characteristics are important factors for designing and construction of steel bridges. Three categories of structural steel are often used for steel bridge construction including carbon steel, high-strength steels, and heat-treated carbon steels (Kumar and Kumar, 2014).

7.2 CONNECTING METHODS

Steel bridges, as well as other steel structures, are built of steel members such as beams, columns, and truss members by connections or joints. The use of connections can affect the fabrication method, serviceability, safety, and the cost, thus they are particularly important in the steel bridge construction. In general, the connection design should follow the principle that should be safe, reliable, simply in design and fabrication, easy installation, and should be able to save the materials and costs.

Bridge Engineering
http://dx.doi.org/10.1016/B978-0-12-804432-2.00007-4

In steel bridges, the often used connecting methods include rivet connection, bolt connections, and welding connections, as shown in Fig. 7.1. Bolt connection is used earliest since the mid-18th century and still is being used as one of the most important connections. The rivet connection has been used since the early 19th century; thereafter the welding connection was also created and used in the end of 19th century. The welding joint became very popular and gradually replaced the rivet connection in the steel bridge construction. With the development of high-strength bolted connection at the mid-20th century, they are also widely used in the steel bridge construction (Wu, 2006).

7.2.1 Bolted Connection

Bolted connection is more frequently used than other connection methods. They are very easy to operate and no special equipment is required. This is in particular due to the development of higher strength bolts, the easy to use

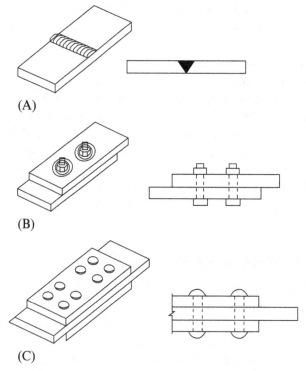

Fig. 7.1 Different connecting methods. (A) Welded connection. (B) Bolted connection. (C) Riveted connection.

and strong structural steel connections become possible. In the bolt design, two kinds of forces including tension and shear forces should be considered. Bolted connection can be divided into ordinary bolted connection or high-strength bolted connection. Both of them are easy in installation, particularly suitable for connection in the construction site. Ordinary bolts are easy to disassemble and are generally used in temporary connections or those need to be disassembled. High-strength bolts are easy to disassemble, and they have higher strength and stiffness. However, the bolted connections also have some disadvantages because it is necessary to drill holes and adjust the holes during the installation. The cutting of the holes may weaken the steel members and increase the use steel materials due to the member overlapping, and also this will increase the workload in the construction. There are many reasons that may result in the failure of the bolted connections, such as overloading, over torquing, or damage due to corrosion.

7.2.2 Rivet Connection

From the mechanical behavior and design points of view, the rivet connection is very similar to ordinary bolt connection. A rivet is a permanent mechanical fastener, which was very popular for the early steel bridges due to their good performance in plasticity, toughness, integrity under statistic load, and fatigue performance under dynamic load. Also, quality inspection of welded connection is also relatively easy than other connection methods.

However, the rivet connection is rarely used in nowadays due to disadvantages like complex in structure, high consumption of steel, high noise during the construction, etc., and gradually replaced by the bolted connection and welded connection.

7.2.3 Welded Connection

Welding is another connecting method used to connect steel components in the fabrication factory and on bridge construction site. Common types of welds are butt welds, fillet welds, and plug welds, as shown in Fig. 7.2.

The work place (in a factory or on site) is an important criterion for deciding whether to choose a bolted or a welded connection. If the connection is performed in a factory, it is generally most economically achieved through welding. Although it is technically possible for site welding, the additional cost for setting up welding and testing facilities as well as the increased erection time usually makes bolted connections become more efficient.

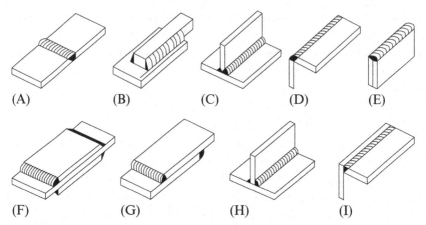

Fig. 7.2 Welded connections. (A) Butt joint. (B) Longitudinal joint. (C) Butt joint. (D) Corner joint-1. (E) Edge joint. (F) Transverse fillet joint. (G) Transverse fillet joint. (H) Tee joint. (I) Corner joint-2.

7.3 STEEL-CONCRETE COMPOSITE BRIDGES

7.3.1 General of Steel-Concrete Composite Girders

The steel bridge is mainly used for large span bridges. With the development of cable-stayed bridges after the world war-II, the use of prestressed concrete became also possible to build long-span bridges. During the competition between steel and concrete bridges, a new type of steel-concrete composite bridge was created. A typical steel-concrete composite section has a steel girder below the concrete deck used for directly supporting the live load, and the shear connection devices are used to ensure the connection between the steel and concrete. Although both concrete and steel are used to build the composite section, as the steel girder is used as the main supporting structure, the steel-concrete composite bridges are generally classified as the steel bridges.

Composite section has been widely used in bridge construction due to the high compressive strength of concrete and high tensile strength of structural steel. As discussed previously in Chapter 6, the compression of the concrete is much stronger than its tension. Although the steel is believed to have similar behaviors in both tension and compression, bucking is always a concern when the structural steel is in compression. For those reasons, the concrete is generally used for compression, while the steel is mainly designed for tension in a composite section. By joining the two materials together structurally these strengths can be used to result in a lightweight and highly efficient design. New composite structures also offer benefits in terms of speed of construction and less use of falsework and formwork.

7.3.2 Shear Connection Devices

7.3.2.1 General

The shear flow or shear stress between steel girder and concrete slab is a natural consequence of the requirement for composite action. If no shear connection, the beam and the concrete slab would behave separately with small load carrying capacity due to the small value of secondary moment of inertia. While the use of shear connectors can enable the two components work together and prevent the slip between them, then it achieves a much stiffer and stronger section.

A simple example is given to explain the difference between the composite and noncomposite beam. As shown in Fig. 7.3, a simply support beam is designed as composite and noncomposite sections, respectively. For the section as shown in Fig. 7.4, the secondary moment of inertia of the composite section is four times that of the noncomposite section, which means the stiffness of the composite section is four times of the noncomposite section. The example clearly shows that the connection on the interface

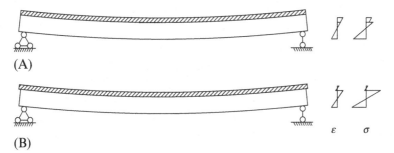

Fig. 7.3 (A) Composite beam. (B) Noncomposite beam.

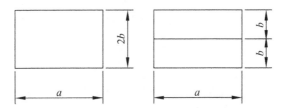

Fig. 7.4 Difference between composite and noncomposite beams.

between the steel girder and concrete deck is the key for the performance of the composite section.

7.3.2.2 Classification of Shear Connectors

The shear connectors used in steel-concrete composite structures can be classified into four types: (1) mechanical joint, (2) friction bonding, (3) adhesion, and (4) adhesives; each of which contains several methods, as shown in Table 7.1. Among these four types, mechanical joints have been widely used because high composite effect can be expected even with small contacting area (JSCE, 2007).

In highway bridges, headed stud connectors have been very widely used due to the facts that they are convenience in construction (welding), low cost, and identical performance in different directions. In railway bridges, shear connectors called block dowels are often used. This type of connector consists of a steel block or horseshoe-shaped plate to which a semicircle rebar is welded. This rebar prevents the uplift of the concrete deck. Other than the headed studs and block dowels, there are perforated plate dowels and angle connector which unifies the flange plate and concrete deck in corrugated steel-web bridges (Igase et al., 2002). Typical shear connectors are shown in Fig. 7.5.

In addition to the mechanical type shear connectors, other connection methods such as high tension bolts are used for precast concrete deck panels. The adhesion type connection method, like the use of rubber-latex mortar, was recently reported to be useful in strengthening aged bridge structures. It is reported that such materials are effective for enhancing the bonding on steel-concrete interface (Lin et al., 2013a, 2014a,b). However, as the adhesion type connector keeps the connection only at the interface between the steel and concrete, the long-term reliability and durability of such shear connectors are still a concern in the design of new composite bridges.

Table 7.1 Classification of Shear Connectors

Type	Method	Example
1	Mechanical joint	Headed stud, shape steel, block dowels, perforated-plate dowels, and angle-connector
2	Friction bonding	High tensile bolt
3	Adhesion	Protruded rolled steels, such as checkered steel plate and rugged-surface H-shaped steel
4	Adhesives	Epoxy resin

Fig. 7.5 Typical shear connectors. *(Photos by Lin and Tainguchi.)*

7.3.2.3 Tests on Shear Connectors

Tests on shear connectors can be generally classified into two types, including push-out tests and beam tests. Push-out tests are basically used as the standard test for determining the performance of shear connectors, which are simple and easy to perform. A typical push-out specimen specified in Eurocode-4 is shown in Fig. 7.6, where the shear connectors are used in T-beams with two concrete blocks. In the push-out test, the load is applied on the T-beam to produce the shear forces on shear connectors. The slip on the steel-concrete can be measured, and applied load-slip relationships can be determined. Therefore, the properties of shear connectors (e.g., stiffness, load carrying capacities, etc.) will be determined for the design. In general, bond at the interface between flanges of the steel beam and the concrete should be prevented by greasing the flange or by other suitable means.

In the beam tests for shear connectors, two concentrated loads are applied symmetrical on the beam to produce the longitudinal shear forces on the steel-concrete interface and shear connectors. The image of the beam tests for shear connectors is shown in Fig. 7.7. The shear forces in the shear connectors keep increasing with the increase of the load, until failure. In

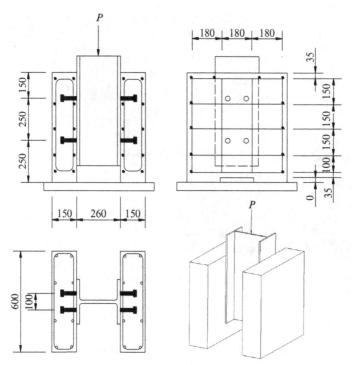

Fig. 7.6 Specimen for push-out test on shear connectors.

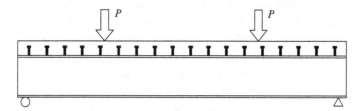

Fig. 7.7 Image of beam test of shear connectors.

comparison with the push–out tests, the beam tests can reflect the real behavior of shear connectors, but rather complicated in practical operation.

Taking the shear stud as an example, the constitutive shear force–slip relationship suggested by Ollgaard et al. (1971) is given by Eq. (7.1) and illustrated in Fig. 7.8. The ultimate shear force carrying capacity specified in JSCE specifications (JSCE, 2007) is shown in Eq. (7.2).

$$Q = Q_u\left(1 - e^{-0.7\Delta}\right)^{0.4} \tag{7.1}$$

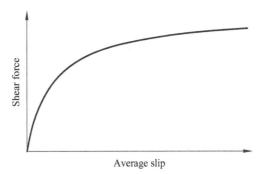

Fig. 7.8 Load-slip relationship of stud.

$$Q_u = \min \left(\frac{\left(31 A_{ss} \sqrt{(h_{ss}/d_{ss}) f_{cd}'} + 10,000 \right) / \gamma_b}{A_{ss} f_{ss} / \gamma_b} \right) \quad (7.2)$$

where Δ is the slip of the shear stud (mm), A_{ss} is the area of the shank of the stud (mm²), d_{ss} is the diameter of the shank of the stud (mm), h_{ss} is the height of the stud (mm), f_{cd}' is the design compressive strength of concrete (N/mm²) $\left(= f_{ck}' / \gamma_c \right)$, f_{ck}' is the characteristic compressive strength of concrete (N/mm²), f_{ss} is the design tensile strength of the stud (N/mm²) $\left(= f_{sk}' / \gamma_s \right)$, f_{sk}' is the characteristic tensile strength of the stud (N/mm²), γ_c is the material factor of concrete (=1.3), γ_s is the material factor of stud (=1.0), and γ_b is the member factor (=1.3).

7.3.2.4 Design of Shear Connectors

1. Elastic design

This method is generally used for rigid or nonductile shear connectors, assuming that the ultimate load is reached when the maximum shear force of any shear connectors becomes equal to its shear resistance. The optimization design of shear connector arrangement is to follow the distribution of the longitudinal shear forces. Therefore, elastic theory is adopted to determine the longitudinal shear per unit length on the interface between the concrete slab and steel girder, as shown below:

$$\tau = \frac{V(x) S}{It} \quad (7.3)$$

where $V(x)$ is the longitudinal shear force at cross section x, S is the first moment of area taken at the steel-concrete interface, I is the second moment of area of the composite section, and t is the width of the top flange of the steel girder.

For unit length, the total longitudinal shear force due to external load should not exceed the shear resistance provided by connectors. Thus, the necessary number N of shear connector at unit length can be determined as:

$$N \geq \frac{\tau t}{P_u} \tag{7.4}$$

2. Plastic design

In the plastic design, the shear connectors are assumed in the ultimate state, and each shear connector sustains the same shear force. In this method, the composite beam is divided into several zones according to the bending moment distribution at the maximum and zero moment points, as shown in Fig. 7.9. For the positive bending moment zones, the necessary shear force is determined as:

$$V = \min \left(A_s f_y, A_c f_c \right) \tag{7.5}$$

For the zones from the maximum positive bending moment and the maximum negative bending moment, the necessary shear force is determined as:

$$V = \min \left(A_s f_y, A_c f_c \right) + A_{st} f_{yt} \tag{7.6}$$

The shear capacity of each shear connector is assumed as P_u, and the required number of shear connectors, N, can be determined by:

$$N \geq \frac{V}{P_u} \tag{7.7}$$

In each zone, the shear connectors are uniformly distributed. In this method, the longitudinal reinforcement in compression is neglected.

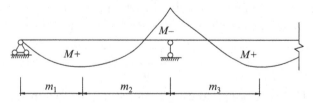

Fig. 7.9 Shear connector division zones in continuous composite beams.

7.3.3 Continuous Composite Beams

7.3.3.1 General

Steel-concrete composite structures have been used extensively in bridge structures, both highway and railway bridges, due to the benefits of combining the two construction materials, their higher span-to-depth ratio, reduced deflections, and higher stiffness ratios than traditional steel or concrete beam structures (Lin et al., 2016a). For simply supported composite beams, the concrete is in compression and the steel is in tension, the composite section is strong to sustain the external loads. For composite beams in the negative moment regions, however, the concrete slab is in tension and the lower flange of the steel beam is in compression which generally has shortcomings in view of the durability, loading capacity, and service life of the structures (Lin et al., 2013b). As shown in Fig. 7.10, for the negative bending moment regions, the concrete cannot resist the tensile stress and therefore cracking occurs, thus only the embedded reinforcement is still effective in resisting moment. Simultaneously, the steel section near the intermediate support is in compression, and the buckling becomes possible and should be considered in the design. The images of the composite girder subjected to negative bending moment in the laboratory test are shown in Fig. 7.11.

In recent years, several researches were performed on the mechanical behavior of composite beams under a hogging moment. Special attention generally was paid to the steel-slab interface (Nie et al., 2004), effective slab width (Aref et al., 2007), crack width control (Ryu et al., 2007), the use of prestress for continuous composite beams (Chen and Jia, 2008; Chen et al., 2009), numerical studies (Korkess et al., 2009), effects of shear connectors (Lin and Yoda, 2011, 2013; Lin et al., 2013b), long-term behavior (Fan et al., 2010a,b), and so on. In contrast to previous studies on the I-girders,

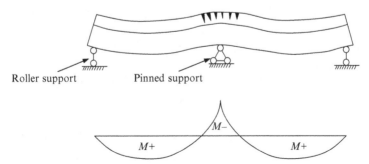

Fig. 7.10 Bending moment distribution and cracking of the concrete slab.

Fig. 7.11 Cracking of concrete slab under negative bending moment. *(Photo by Lin.)*

the mechanical behavior of composite box-girders under a hogging moment was also studied recently (Su et al., 2012). But in the bridge design, the following methods are generally used for improving the performance of the composite section subjected to negative bending moment.

7.3.3.2 Improvement of Composite Section Subjected to Negative Moment

1. Concrete pouring sequence

    In the composite bridge construction, it is possible to improve the performance of the composite section by changing the construction sequence. The basic concept is shown in Fig. 7.12: the concrete was first poured, and addition temporary loads are also applied in the positive bending moment region, thus "prestress" will be produced in the negative bending moment region. Thereafter, the concrete is poured in the negative bending moment, and the temporary loads will not be removed before the hardening of the concrete in the negative bending moment region. After the removal of the additional temporary loads, the concrete in hogging bending moment region will be under compression. The initial compressive stress in concrete will be used for reducing the cracking of the concrete in the service stage.

2. Bearing adjustment

    Bearing adjustment is another way to improving the composite section under negative bending moment region. As shown in Fig. 7.13, certain upward movement is applied on the intermediate bearings before

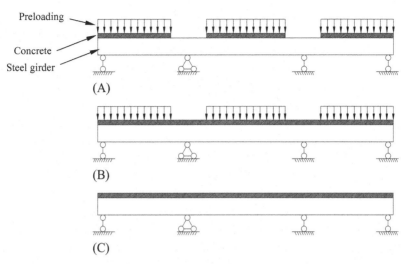

Fig. 7.12 Improvement of the negative bending moment zone by construction sequence and preloading. (A) Concrete casting and preloading in positive bending moment region. (B) Concrete casting in negative bending moment region. (C) Removal of the preloading.

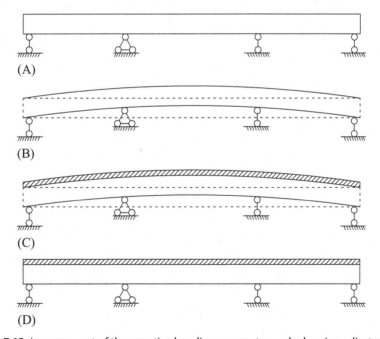

Fig. 7.13 Improvement of the negative bending moment zone by bearing adjustment. (A) Steel girder erection. (B) Upward of the intermediate bearings. (C) Concrete casting. (D) Downward of the intermediate bearings.

the concrete pouring. After the concrete hardening, the intermediate bearings will be moved downward to the designed position. By doing this, a certain level of "prestress" is applied to concrete and the cracking of the concrete is possibly to be avoided. Different from the previous method, the "prestress" in this method is applied to the whole bridge.

3. Prestressing

It can be found that the basic idea of previous two methods is to apply the prestress on the concrete before it is subjected to the service load the possible tensile stress. In the third method, therefore, the prestressing is directly applied to the composite section in the negative bending moment region. Prestressing can be applied by using either prestressing bars or prestressing tendons. The prestressing bars can be directly used in the negative bending moment to apply the compressive stress on the concrete and to avoid the cracking of the concrete. This method can apply the compressive stress only in the local negative bending moment region and do not have to affect other regions. Or instead, the prestressing can be applied by using the prestressing tensions used in the whole bridge. This method can improve the stress distribution in both the concrete and steel girders. But it should be noted that suitable position of the prestressing tendons should be designed in order to achieve the best stress, and an example is shown in Fig. 7.14. In addition, the prestress loss in both construction and service stages would be a concern in its practical use.

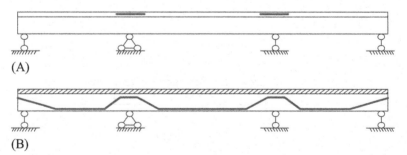

Fig. 7.14 Improvement of the negative bending moment zone by prestressing. (A) Use of prestressing bar. (B) Use of prestressing tendon.

7.4 CASE STUDY—A RESEARCH ON STEEL—CONCRETE COMPOSITE BEAMS SUBJECTED TO HOGGING MOMENT

A series of experimental studies have been performed at Waseda University financed by the Ministry of Land, Infrastructure, Transport and Tourism (MLIT) and Japan Society for the Promotion of Science (JSPS, Grant-in-Aid for Young Scientists B (15K18108)). This study examines experimentally the behavior of composite steel-concrete beams. A number of specimens were tested to study the various aspects of composite beams under negative bending moment, including the effects of repeated load level, shear connectors (PBL and Stud), rubber-latex mortar, and steel fiber reinforced concrete (SFRC) on their structural performance.

7.4.1 Experimental Program

Ten simply supported steel-concrete composite beams were designed and constructed in this study. Four specimens were used to perform the fatigue test under different repeated load levels, and the other six specimens were tested under static load to study the various aspects of composite beams under negative bending moment, including the effects of shear connectors (PBL and Stud), rubber-latex mortar, and SFRC on their structural performance. All specimens were tested to ultimate stage by being subjected to negative moments applied over their entire spans, as simply supported beams loaded at mid-span. This test set-up approximately simulated a portion of a two-span continuous beam between the inflection points on the two sides of an intermediate support, as shown in Fig. 7.15E and F. The approximate simulation was due to the linear bending moment diagram for the tested beams, compared with the nonlinear moment diagram for a continuous beam (Ayyub et al., 1992).

7.4.2 Test Specimens

Each of the specimens was 4600 mm in length and was simply supported at a span of 4000 mm. The concrete slab thickness was 250 mm with a width of 800 mm. Vertical stiffeners were welded at supports, loading points to prevent shear buckling failure and crippling of the web before flexural failure. The transverse reinforcements with a nominal diameter of 13 mm and longitudinal reinforcements with a nominal diameter of 19 mm were arranged on both the top and the bottom of the concrete slab. The longitudinal

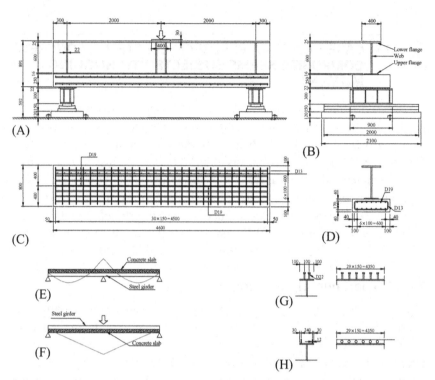

Fig. 7.15 Dimensions of test specimen. (A) Front elevation. (B) Side elevation. (C) Reinforcing bars. (D) Sectional view. (E) Negative moment region in continuous composite girder. (F) Overturned simply supported composite girder. (G) Stud. (H) PBL.

reinforcement ratios were 1.98% for all the specimens. The typical geometry of test specimen is shown in Fig. 7.15, and the details are summarized in Table 7.2.

Ten steel-concrete specimens in total were used in this experimental study. Specimen-1 and -2, with normal concrete, were tested to study the fundamental behavior of the composite beams subjected to hogging moment. Specimen-3 and -4, with headed stud as shear connectors, were tested under different levels of repeated load (initial cracking or stabilized cracking load) to study the fatigue behavior under real service load. Specimen-5 (with normal concrete) and -6 (with SFRC) were tested under different levels of repeated load to study the fatigue behavior of the specimens with PBL as shear connectors. Specimen-7 and -8 with SFRC were tested to confirm its resistance to cracking and crack propagation. The fiber reinforcement is in the form of short discrete fibers with the length of 20–60 mm and the diameter of 0.2–0.6 mm, and the swelling agent was used

Table 7.2 Details of Test Specimen

		Design Parameters		
Specimen	**Connector**	**Load Type**	**Concrete (N/mm^2)**	**Rubber-Latex Coating**
Specimen-1	Stud	Static	Common	No
Specimen-2	PBL	Static	Common	No
Specimen-3	Stud	Initial fatigue	Common	No
Specimen-4	Stud	Stationary fatigue	Common	No
Specimen-5	PBL	Initial fatigue	Common	No
Specimen-6	PBL	Stationary fatigue	Steel fiber + swelling agent	No
Specimen-7	Stud	Static	Steel fiber + swelling agent	No
Specimen-8	PBL	Static	Steel fiber + swelling agent	No
Specimen-9	Stud	Static	Common	Yes
Specimen-10	PBL	Static	Common	Yes

Note: Initial fatigue, repeated load is equivalent to the initial cracking load of 200 kN, 2×10^6 load cycles; Stationary fatigue, repeated load is equivalent to the stabilized cracking load of 600 kN, 2×10^6 load cycles.

in the SFRC. And finally, Specimen-9 and -10 were tested to study the effect of rubber-latex mortar coating on the structural behavior of the composite beams (Fig. 7.16).

7.4.3 Instrumentation and Testing Procedure

The beams were instrumented for the purpose of measuring deflections, crack formation and crack width development, sectional strains across the depth, applied load, and slip between the steel beam and concrete slab. The deflection at mid-span was measured by using two linear variable displacement transducers (LVDTs). The deflections at both ends were also recorded. Another 14 LVDTs were employed along the longitudinal direction to measure the slip on the interface between the concrete slab and the steel beam. Strain gauges used to measure the strain of the steel beam, concrete slab, reinforcing bars were also employed. Besides, 14 π-gauges were arranged on the top surface of the concrete slab to measure the cracking formation and the crack width development. To simulate the negative moment region over the support, between the inflection points of a continuous bridge, the beams were inverted such that the concrete slab was at the

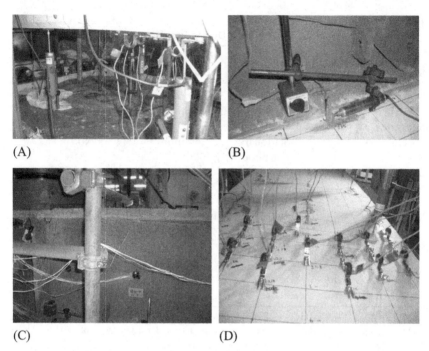

(A)   (B)

(C)   (D)

Fig. 7.16 Arrangement of LVDTs, π gauges, and strain gauges in the test. (A) LVDTs for vertical displacement. (B) LVDTs for slip measurement. (C) Strain gauges. (D) π gauges.

bottom resting on the supporting edges. The beams were tested by applying a single concentrated load at mid-span with roller (simple) end supports. The concentrated load was applied from the top at mid-span of the specimen representing the support of an actual bridge.

7.4.3.1 Static Loading Test

The test specimen was supported by a roller system at two ends, and the static loading test set-up is shown in Fig. 7.17A. After the drying shrinkage had stabilized, preloading was applied to check the reliability of the measure equipment and the stability of the test specimen. The negative bending load was conducted by static loading with unloading process once at the levels 200, 400, 700, and 1300 kN with loading rates of 0.001, 0.0015, 0.002, and 0.003 mm/s, respectively. Displacement control was used in the tests with a loading rate of 0.004 mm/s for the subsequent experiment. For unloading process in the experiments, the load was removed at the calculated initial cracking load and at the stabilized cracking load to check the

Fig. 7.17 Static test, fatigue test, and impact test set-up. (A) Static loading test. (B) Fatigue loading test. (C) Impact test.

cracking of the concrete slab. The loading was terminated when either the maximum stroke of the jack was reached or when the loading capacity of the test specimen dropped significantly.

7.4.3.2 Fatigue Loading Test

The fatigue set-up is shown in Fig. 7.17B, and the repeated cyclic load was applied by a large-scale fatigue testing machine at a frequency of 2–3 cycles per second. The repeated load P_r was determined using the initial cracking load P_{cr} and stabilized cracking load P_s. The absolute maximum loads applied to the specimens were 200 kN for Specimen-3 and Specimen-5, and 600 kN for Specimen-4 and Specimen-6 separately. Before the fatigue test, the initial static test with the load up to 200 kN was performed. During the fatigue test, the repeated loading was periodically stopped after typical load cycles, such as 1×10^4, 5×10^4, 1×10^5, 5×10^5, 1×10^6, and 2×10^6, to conduct intermediate static tests for determining the structural performance of the specimen. Sectional strains, crack widths, deflections, and reactions were measured. Similar to the repeated load in the fatigue test, the maximum static load was also limited to the theoretical initial cracking (200 kN, Specimen-3 and Specimen-5) or stabilized cracking load (600 kN, Specimen-4 and Specimen-6).

7.4.3.3 Impact Test

In order to evaluate the noise reduction effect of rubber-latex mortar, the vibration and the sound pressure were measured test specimens. Vibration accelerations were recorded by using accelerometers arranged on the web. Also, microphones were used to measure the noise levels at different distance from the bridge. The arrangement of the accelerometers and microphones was shown in Fig. 7.17C.

7.4.4 Experimental Results and Main Findings

7.4.4.1 Effect of Shear Connectors

Studs and perfo-bond strips (PBLs), generally known as flexible and rigid shear connectors separately, are most frequently used shear connectors in composite girders. However, very few studies were performed particularly to investigate the effects of different shear connectors on mechanical performance of composite beams subjected to negative bending moment. For this reason, experimental tests were performed in this study (Fig. 7.18) to investigate the difference between studs and PBLs on inelastic mechanical behavior of composite girders subjected to negative bending moment.

The results obtained in this study indicate that the PBL specimens have relative large beam stiffness, ultimate load carrying capacity but relative small slip on the steel–concrete interface. As the cracking of the concrete slab, the composite neutral axis moves away from the PBL plates, resulting in more effective of PBL plates in contribution to beam rigidity and load carrying

Fig. 7.18 Stud and PBL shear connectors.

capacity in comparison with stud shear connectors. The interaction between PBL connectors, reinforcing bars (transverse bars go through the holes of the PBL), and the concrete restricts the interface slip. Besides, the results also indicate that the Stud specimens have better mechanical behavior in regard to concrete slab cracking behavior, such as initial cracking and crack closure. But all in all, both PBLs and stud connectors are effective shear connective devices for composite beams subjected to negative bending moment.

7.4.4.2 Effects of SFRC

SFRC is a structural material characterized by a significant residual tensile strength in postcracking region and enhanced capacity to absorb strain energy due to fiber bridging mechanisms across the crack surface. For improving long-term behavior, enhancing strength, toughness, and stress resistance, SFRC is being used in structures such as flooring, housing, precast, bridges, tunneling, heavy duty pavement, and mining. There has been considerable research on the mechanical behavior of SFRC in recent years. The past research results indicate that the SFRC has relatively good properties in comparison with the normal concrete. Therefore, the use of SFRC could be a solution for improving the mechanical performance of the composite steel-concrete beams under negative bending moment.

This study investigates the effects of SFRC on improving the mechanical performance of composite steel and concrete beams subjected to hogging moment. SFRC (Fig. 7.19) was used for three specimens, and normal concrete was applied for the other specimens. Shrinkage of the normal and SFRC, load versus mid-span deflection relationship, crack formation and its propagation process, and slip development on the steel-slab interface were measured and investigated. The test results show that the inclusion of steel fibers decreased both the crack spacing and crack width of the concrete slab. The specimens with SFRC have relatively large initial cracking load, and crack width of the concrete slab can be controlled appropriately in the service stage. The effects of the SFRC on the ultimate load carrying capacity of the beam depend on the shear connectors applied on the steel-slab interface. The application of SFRC could enhance the load carrying capacity of the specimen with stud shear connectors; however, the effects of SFRC on the load carrying capacity of PBL specimen were found to be declined in some extent.

7.4.4.3 Effects of Repeated Load

In the real service condition, bridge structures are generally subjected to vibrating or oscillating forces. The behavior of structures under such loading

(A)

(B)

Fig. 7.19 Specimens with steel fiber reinforced concrete (SFRC). (A) Sketch of the steel fibers. (B) Mixing of steel fibers and concrete.

conditions differs from the behavior under a static load. Because the struc-ture member is subjected to repeated load cycles (fatigue) in actual use, designers are faced with predicting fatigue life or its effect on structural mechanical behavior of structures. As a result, fatigue testing generally gives much better data than common static loading test to predict the in-service life of bridge structures. On this background, this study deals with the results of a series of experimental work with three test steel-concrete composite specimens. This study keeps a watchful eye on the problems involving dif-ferent repeated load levels on the girder stiffness, crack formation and prop-agation on the surface of the concrete slab, strain development process of reinforcing bars, shear studs and concrete interaction, and flexural strain change process of studs as well as fatigue behavior of the concrete slab. Besides, a brief description about the final static tests that followed by the fatigue tests was made to clarify the influence of repeated loading on static loading behavior of such beams.

Test results indicated that when the repeated load was equivalent to the initial cracking load, the fatigue test had only limited influence on beam stiffness or crack patterns. However, when the repeated load was equivalent to the stabilized cracking load, a number of residual cracks occurred in the initial static test and the beam became less stiff as the load cycles increase. Failure of the bond between studs and surrounding concrete was confirmed in intermediate static tests; flexural stiffness of studs became smaller as the increase of load cycles. In addition, final static tests were performed on fatigue test specimens and another specimen without fatigue test. The comparison indicated that when the repeated load was larger than the initial cracking load, fatigue test could decrease beam stiffness and its ultimate load carrying capacity.

7.4.4.4 Effects of Rubber-Latex Mortar Coating

Concrete and mortar, including SBR latex, shows various abilities especially in adhesion bonding, waterproofing, shock absorption, and abrasion resistance. The rubber-latex mortar was used for steel structures to reduce the noise as well as the interface slip, enhance the integrity, and improve the durability of the composite structures. Moreover, the composite structures with rubber-latex mortar are reported to have the good impact resistance, weather resistance as well as the good constructability and operability behaviors. In order to study the effect of rubber-latex mortar coating on structural behavior of composite girders under negative bending moment, rubber-latex mortar is sprayed on two specimens. The image of the rubber-latex mortar spraying shown can be seen in Fig. 7.20. Both static loading test and impact tests were performed, and the test results of the specimens with rubber-latex mortar were compared with those without using rubber-latex mortar.

The rubber-latex mortar-coated composite beams were designed to offer corrosion resistance, high initial cracking loading capacity in combination with higher bond-transfer characteristics on the steel-slab interface than the basic uncoated composite beam. According to the test results, the composite beams with rubber-latex mortar coated on the surface of the steel beam, shear connectors, and the concrete slab showed better initial cracking strength than that of those without rubber-latex mortar. Besides, adhesion bonding effects of rubber-latex on steel-slab interface and interface between the shear studs and surrounding concrete were also confirmed in the test.

Fig. 7.20 Rubber-latex mortar coating.

In addition, the impact test results indicate that the noise levels can be greatly reduced by using rubber–latex.

In addition to the studies performed for composite beams subjected to pure negative bending moment, the mechanical performance of composite beams under combined negative bending and torsional moments (Fig. 7.21) were also performed recently (Lin et al., 2016b). Many interesting results are expected to be obtained in the near future.

Fig. 7.21 Steel-concrete composite beams under combined negative bending and torsional moments.

7.5 SUMMARY

With the development of automated fabrication and new construction techniques, steel bridges are able to provide economic solutions to the demands of safety, long-span capacity, rapid construction, esthetic appearance, minimal maintenance, and flexibility in future use. As the structural steel has a high strength-to-weight ratio, the steel bridges can be built from very short span to very long span, supporting the imposed loads with the minimum of deadweight. However, the fatigue and corrosion problems of steel should be considered in the whole service life of the bridge.

To make full use of the material strength of steel and concrete, the steel-concrete composite bridges are often built. For continuous steel-concrete composite bridges, the cracking of concrete can be avoided by several means. As a typical type of steel bridges, the truss bridge will be discussed in Chapter 8.

7.6 EXERCISES

1. Describe the connection methods for steel members.
2. Describe the conception idea and advantages of steel-concrete composite bridges in comparison with traditional steel bridges or concrete bridges.
3. Describe the function of shear connectors in composite structures and their classifications.
4. Describe the problems related to steel-concrete composite beams subjected to negative bending moment and possible improving methods.

REFERENCES

Aref, A.J., Chiewanichakorn, M., Chen, S.S., Ahn, I.S., 2007. Effective slab width definition for negative moment regions of composite bridges. J. Bridge Eng. 12 (3), 339–349, 10.1061/(ASCE)1084-0702.

Ayyub, B.M., Sohn, Y.G., Saadatmanesh, H., 1992. Prestressed composite girders. I: experimental study for negative moment. J. Struct. Eng. ASCE 118 (10), 2743–2762.

Chen, S., Jia, Y., 2008. Required and available moment redistribution of continuous steel—concrete composite beams. J. Constr. Steel Res. 64 (2), 167–175.

Chen, S., Jia, Y., Wang, X., 2009. Experimental study of moment redistribution and load carrying capacity of externally prestressed continuous composite beams. Struct. Eng. Mech. 31 (5), 605–619.

Fan, J., Nie, J., Li, Q., Wang, H., 2010a. Long-term behavior of composite beams under positive and negative bending. I: experimental study. J. Struct. Eng. ASCE 136 (7), 849–857.

Fan, J., Nie, J., Li, Q., Wang, H., 2010b. Long-term behavior of composite beams under positive and negative bending. II: analytical study. J. Struct. Eng. ASCE 136 (7), 858–865.

Igase, Y., Suzuki, N., Suzuki, T., Yanaka, T., 2002. An experimental study on horizontal shear slip characteristics and fatigue strength of angle-connector. In: Proc. of Annual Conf. of JSCE, I, vol. 57, pp. 725–726.

Japan Society of Civil Engineers, 2007. Standard Specifications for Steel and Composite Structures, first ed. Japan Society of Civil Engineers, Tokyo

Korkess, I.N., Yousifany, A.H., Majeed, Q.A., Husain, H.M., 2009. Behavior of composite steel-concrete beam subjected to negative bending. Eng. Technol. J. 27 (1), 53–71.

Kumar, S.R.S., Kumar, A.R.S., 2014. Design of steel structures. Indian Institute of Technology Madras. Available from http://nptel.ac.in/courses/105106113/56.

Lin, W., Yoda, T., 2011. Mechanical behaviour of composite girders subjected to hogging moment: experimental study. J. Jpn. Soc. Civ. Eng. Ser A1 (Struct. Eng. Earthq. Eng.) 67 (3), 583–596.

Lin, W., Yoda, T., 2013. Experimental and numerical study on mechanical behavior of composite girders under hogging moment. Adv. Steel Constr. Int. J. 9 (4), 280–304.

Lin, W., Yoda, T., Taniguchi, N., Hansaka, M., 2013a. Performance of strengthened hybrid structures renovated from old railway steel bridges. J. Constr. Steel Res. 85, 130–139.

Lin, W., Yoda, T., Taniguchi, N., Kasano, H., He, J., 2013b. Mechanical performance of steel-concrete composite beams subjected to a hogging moment. J. Struct. Eng. ASCE 140 (1), 04013031.

Lin, W., Yoda, T., Taniguchi, N., 2014a. Rehabilitation and restoration of old steel railway bridges: laboratory experiment and field test. J. Bridg. Eng. ASCE 19 (5). 04014004.

Lin, W., Yoda, T., Taniguchi, N., Satake, S., Kasano, H., 2014b. Preventive maintenance on welded connection joints in aged steel railway bridges. J. Constr. Steel Res. 92, 46–54.

Lin, W., Yoda, T., Taniguchi, N., Lam, H., Nakabayashi, K., 2016a. Post-fracture redundancy evaluation of a Twin Box-Girder Shinkansen Bridge in Japan. In: Proceeding of IABSE Conference Guangzhou.

Lin, W., Lam, H., Yoda, T., Middleton, C.R., 2016b. Steel-concrete composite beams subjected to combined hogging bending and torsion elastic behavior of steel-concrete composite beams under negative bending moment. In: Eighth International Conference on Steel and Aluminum Structures, Hong Kong, China.

Nie, J., Fan, J., Cai, C.S., 2004. Stiffness and deflection of steel-concrete composite beams under negative bending. J. Struct. Eng. ASCE 130 (11), 1842–1851.

Ollgaard, J.G., Slutter, R.G., Fisher, J.W., 1971. Shear strength of stud connectors in lightweight and normal weight concrete. Eng. J. AISC 8, 55–64.

Ryu, H.K., Kim, Y.J., Chang, S.P., 2007. Crack control of a continuous composite two-girder bridge with prefabricated slabs under static and fatigue loads. Eng. Struct. 29 (6), 851–864.

Su, Q., Yang, G., Wu, C., 2012. Experimental investigation on inelastic behavior of composite box girder under negative moment. Int. J. Steel Struct. 12 (1), 71–84.

Wu, C., 2006. Modern Steel Bridges. China Communications Press, Beijing.

CHAPTER EIGHT

Truss Bridges

8.1 INTRODUCTION

A truss acts like a beam but with its members subjected primarily to axial forces. In mechanics, a truss is generally defined as a structure with connected two-force (either tension or compression) members forming triangular units. The members are arranged in triangular patterns, and the forces are applied only at both end points and adjacent members are connected at joints referred to as connections. For members in a truss, therefore, bending and torsion moments are generally explicitly excluded because all the joints are assumed that only axial forces can be resisted and the joints have no rotation freedoms.

Following this definition, a bridge whose load carrying superstructure is composed of a truss is called the truss bridge. The truss elements may be in tension, compression, or sometimes either in tension or compression in response to dynamic loads. From a mechanical point of view, truss structures are highly efficient in using the strengths of construction materials due to the fact that only axial forces are resisted in truss members.

The truss bridge is also called a beam bridge with braces. A comparison between the sectional stress distribution between a beam and a truss is shown in Fig. 8.1. For a beam, the top surface of the beam gets the most compression, and the very bottom of the beam experiences the most tension, while the middle of the beam (near the neutral axis) experiences very little compression or tension. Thus it is more reasonable to provide more material on both top and bottom of beams to better handle the forces of compression and tension. Therefore, I-shaped beams and trusses are good options under these circumstances. Top and bottom flange members of a truss are designed to sustain the sectional bending moment, while the diagonal members (or web systems) are designed for sustaining the shear forces, resulting in the effective use of materials. Truss bridges are one of the oldest types of modern

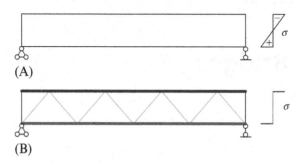

Fig. 8.1 (A) Beam. (B) Truss.

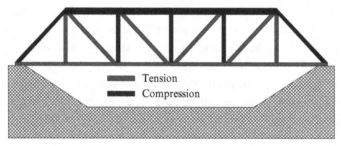

Fig. 8.2 Axial forces in a truss.

bridges. Moreover, a truss is generally more rigid than a beam because a truss is composed of a variant of triangles and it has the ability to dissipate a load through the whole truss. An example is shown in Fig. 8.2.

8.2 TRUSS BRIDGE TERMINOLOGY

Principal parts and terminologies of a truss bridge are shown in Figs. 8.3 and 8.4, respectively.

8.2.1 Joints or Connections

Joints, also referred to as connections, are intersections of truss members. Joints are one of the most important components, as their damage or failure renders connected members no longer functional, resulting in the partial or complete collapse of the bridge. In the truss bridge design, live loads are generally transmitted through floor framing to the panel points of either chord in order to minimize bending stresses in truss members.

There are two types of joints, namely pinned connections and gusset plate connections (Fig. 8.5), often used in truss bridges. The pinned

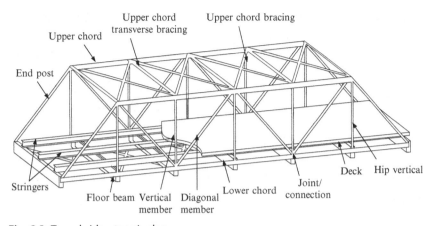

Fig. 8.3 Truss bridge terminology.

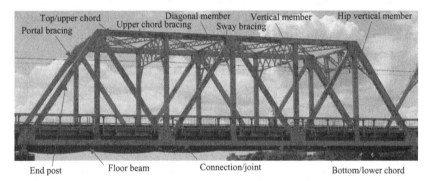

Fig. 8.4 Terminology of truss bridges. *(Photo by Lin.)*

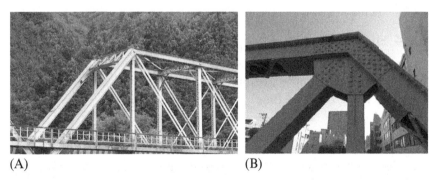

(A)   (B)

Fig. 8.5 Truss connections. (A) Pinned connection. (B) Gusset plate connections. *(Photo by Taniguchi and Lin.)*

connection uses a single large metal pin (or hinge) to connect two or more members together, which are mainly used in old truss bridges. In modern truss bridges, truss members are generally connected by one or two heavy metal gusset plates, called gusset plate connection. The use of connections also affects the design methodology (or structural analyses) of truss bridges. Truss bridges are generally considered as statically determinate structures. For pinned connections, this assumption is closer to the truth due to the free rotation in the pin, and the bending stress can be ignored. For modern truss bridges with gusset plate connections, however, the structures are statically indeterminate from the mechanical point of view, and the stress analysis is difficult.

8.2.2 Chord Members
Chord members include both top chord (also referred to as upper chord) and bottom chord (also referred to as lower chord), and they act like the flanges of a beam. They resist the tensile and compressive forces caused by sectional bending moment. The bottom chords are generally parallel. The top chords are essentially parallel in a truss bridge with constant depth, but they may also be designed as inclined members in a truss bridge with variable-depth, such as in continuous trusses.

8.2.3 Web Members
Web members consist of diagonals and verticals. The diagonals carry the shear forces in the truss sections. The verticals also provide both shear force capacities and additional loading supports for reducing the span of the chords.

8.2.4 End Posts
End post is the outermost vertical or diagonal member of a truss bridge, they need to sustain the compressive forces from the dead and live loads and transmit those loads to the bearings.

8.2.5 Deck
The deck in the truss bridge is similar to those in other bridge types. They directly support loads (vehicular loads or pedestrian loads etc.) and transmit them to the floor beams.

8.2.6 Floor Beams

Floor beams are transverse to the traffic direction connecting two trusses, and they are designed to support the loads from the stringers and transmit them to the main trusses.

8.2.7 Stringers

Stringers are longitudinal beams under the deck, set parallel to the traffic direction. They support the deck load directly and transmit them to the floor beams. In case of a truss bridge without stringers, the deck is designed to be supported directly by the floor beams.

8.2.8 Lateral Bracings

Lateral bracings are members connecting the chords to each other, and they are mainly designed to ensure stability of the truss and to provide lateral resistance to wind.

8.2.9 Sway Bracings

Sway bracings are members connecting the top chords to each other, and they are only used in through truss bridges. Sway bracings should be carefully designed to provide adequate clearance for the traffic below it.

8.2.10 Portal Bracings

Portal bracing is sway bracing connecting the top of the end posts.

8.3 TYPES OF TRUSSES

8.3.1 According to Structural Systems

There are three common truss configurations that are often used in bridges, namely Warren truss, modified Warren truss, and Pratt truss, as shown in Fig. 8.6. All these truss configurations can be used as an underslung truss, a semithrough truss, or a through truss bridge.

Warren trusses have parallel chords and alternating diagonals, as shown in Fig. 8.6A. Warren trusses with verticals to reduce panel size are named as modified Warren truss, as shown in Fig. 8.6B. Pratt trusses have diagonals sloping downward toward the center and parallel chords, as shown in Fig. 8.6C.

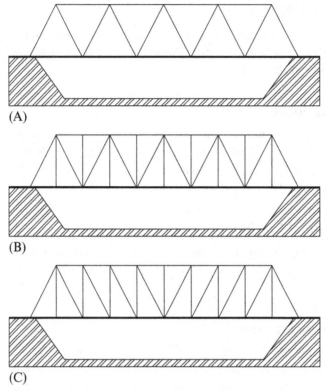

Fig. 8.6 Truss types according to structural forms. (A) Warren truss. (B) Modified Warren truss. (C) Pratt truss.

In addition to the three basic truss systems, there are many other structural forms developed in the bridge design practice, as shown in Fig. 8.7. Some pictures of truss bridges are shown in Fig. 8.8.

8.3.2 According to Deck Locations

A truss bridge may carry its deck on the top, in the middle, or at the bottom of the truss, and they are accordingly classified into deck truss, half-through truss, or through truss, respectively, as shown in Fig. 8.9. When the bridge deck is at top the truss it is called a deck truss, so that vehicles or other live loads are carried above the top chords. The deck in through trusses is placed near the bottom chord so that vehicles pass through the trusses. According to above definition, it can found all the truss bridges shown in Fig. 8.8 are through truss bridges. One deck truss bridge and one half-through truss bridge are shown in Fig. 8.10A and B, respectively. The choice of deck,

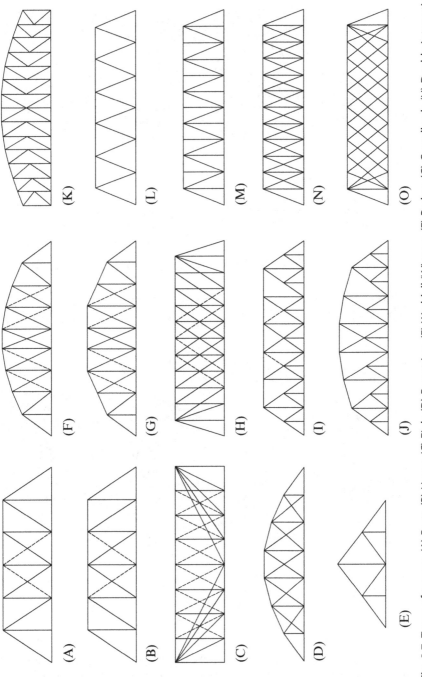

Fig. 8.7 Types of trusses. (A) Pratt. (B) Howe. (C) Fink. (D) Bowstring. (E) Waddell "A" truss. (F) Parker. (G) Camelback. (H) Double interaction Pratt. (I) Baltimore. (J) Pennsylvania. (K) K-truss. (L) Warren. (M) Modified Warren. (N) Double interaction Warren. (O) Lattice.

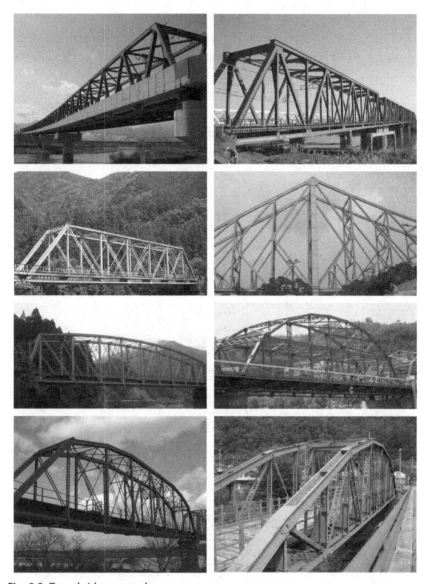

Fig. 8.8 Truss bridge examples.

half-through, or through trusses is normally decided according to the span length, live load types, economical efficiency, etc.

In the design of half-through trusses, however, the absence of top bracing requires special provisions to resist lateral forces. In general, the top chord of a half-through truss bridge is designed as a column with elastic

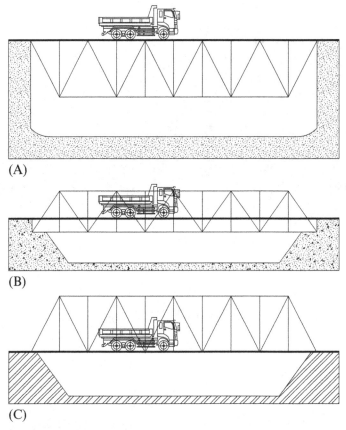

Fig. 8.9 Truss types according to deck location. (A) Deck truss. (B) Half-through truss. (C) Through truss.

lateral supports at truss connections. The critical buckling load of the top chord should be examined so that its stability under design load (considering the impact factor for live load) can be guaranteed.

8.4 THE DESIGN OF TRUSS BRIDGES

Truss members mainly work in tension and compression, thus it can use materials efficiently. The truss bridges can provide large carrying capacity for relatively low consumption of structural steel. In the design of truss bridges, the span numbers, truss height to span length ratio, panel numbers and panel length, and design of member sections deserve special attention.

(A)

(B)

Fig. 8.10 (A) A deck truss bridge. (B) A half-through truss bridge. *((A) The Miyagawa Bridge, photo by Yoda and (B) photo by Lin.)*

8.4.1 Span Numbers

The span number should be decided by considering both superstructure and substructures, this is appropriate for any type of bridge, but particularly useful for design of truss bridges. Less span number means less number of piers but longer span length, thus the reduction of the cost for substructures generally causes increase in cost of superstructures. For truss bridges, it is suggested that the cost of one pier should equal the cost of one superstructure span, excluding the floor system (Brockenbrough and Merritt, 1999).

8.4.2 Height (or Depth) of the Truss

When the span length is decided, the height of the truss can be determined by taking an appropriate depth to span ratio $\left(\dfrac{H}{L}\right)$. The depth of the truss shall also be taken according to the railway or road way clearances, but in general,

$$\left(\frac{H}{L} = \frac{1}{6} \sim \frac{1}{8}\right) \tag{8.1}$$

The depth to span ratio shall be a larger value for a longer span, while a smaller value shall be taken for continuous trusses.

8.4.3 Panel Dimensions

Truss bridges are generally designed as symmetric with even number of panels, and the slope of the diagonals is generally taken between 45 and 60 degrees with the horizontal for economic purpose. The panel length is generally taken from 6 to 10 m. In general, for truss bridge with longer spans, the height as well as the panel length shall be considerably larger. However, as the panel length increases, larger cross section shall be used for compression members like bottom chord and the stringers, resulting in a substantial increase in the cost of the floor system. A subdivided truss can be considered as a possible choice in this occasion.

8.4.4 Design of Truss Members

For modern truss bridges, gusset plate connection (not pinned connection) is generally used. The dimensions of the truss members can be smaller due to the rigid connection between truss members. In general, the height of the truss members can be taken as the 1/15 or 1/20 of the panel length (Tachibana, 2000). Typical cross section for truss members are I-section and box-section depending on the member locations and internal force conditions of truss members, as shown in Fig. 8.11. Box-sections have obvious advantages like ease of maintenance but I-sections are more cost effective. In general, box-section is mainly used for chord member (both upper chord and lower chord), and I-section is mainly used for diagonals in tension.

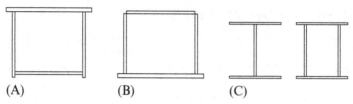

Fig. 8.11 Typical cross section for truss members. (A) Upper chord. (B) Lower chord. (C) Diagonal member.

For chord members, the cross section changes (by changing the member thickness) with the sectional forces, but its height shall remain a constant value to ensure a stabilized gravity center of truss members, which will also benefit the construction at connections.

In truss bridges, the tension members can be designed as compact as possible, but sufficient depths are necessary for providing spaces for installation of the lateral beam and bolts at gusset plates etc. For compression members, the possible buckling problems shall always be considered. Though the effective length for in-plane buckling of compression chord is generally not the same as that for buckling at out-of-plane, the ideal compression chord will be one that has a section with the same slenderness ratio at both planes.

8.5 CASE STUDY—TOKYO GATE BRIDGE

Tokyo Gate Bridge is a 2618-m bridge that spreads across East Tokyo Passage on Tokyo Port Seaside Road linking Jonanjima, Ota ward, and Wakasu, Koto Ward.

This bridge has a unique shape due to the following reasons: (1) Building height restricted due to its proximity to Haneda Airport; and (2) sufficient maximum-bridge-deck height required to ensure the safety of vessels navigating the main channel of Tokyo Port. Therefore, the Bridge employs "truss structure" instead of taking the form of a suspended or cable-stayed bridge that requires a high bridge tower, thereby ensuring its safety and durability to allow for traffic that amounts to approximately 320,000 vehicles a day, including logistic ones.

The construction was carried out from 2002 to 2011, and the bridge was opened to traffic on Feb. 12, 2012 with a total construction cost of approximate 1.1 billion yen. Images of the bridge under construction and whole bridge after construction are shown in Figs. 8.12 and 8.13, respectively. The size dimensions of Tokyo Gate Bridge is illustrated in Fig. 8.14.

Fig. 8.12 Tokyo gate bridge during construction.

Fig. 8.13 Tokyo gate bridge after construction.

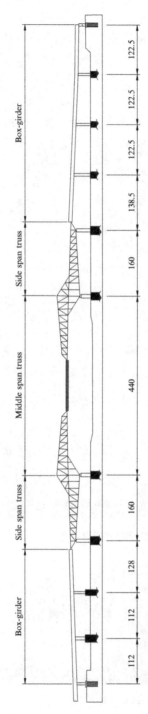

Fig. 8.14 Size dimensions of Tokyo gate bridge (unit: m).

8.6 EXERCISES

1. Explain truss bridge terminology and the different functions of each member.
2. Describe the live load transfer path in a typical truss bridge, and explain why only load was applied on joint when analyzing the main truss.
3. Describe the two common connections in truss bridges and explain the mechanical differences between them.
4. Classify the truss bridges according to the relative location of the deck and the main truss, and explain their differences in design.
5. Prepare a truss model and perform a loading test on it as shown in Fig. 8.15 according to the following requirements:
   (a) One student needs to build one model, and no restrictions in selecting truss forms;
   (b) Materials can be used: toothpicks (maximum number: ≤ 50) and glue stick only;

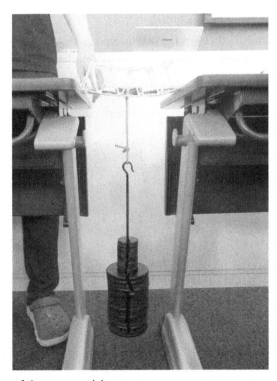

Fig. 8.15 Image of the truss model test.

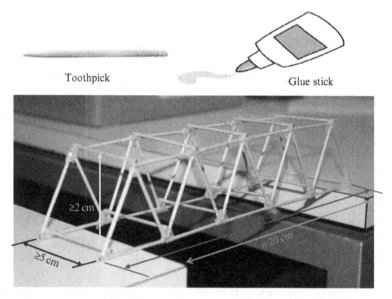

Toothpick   Glue stick

Fig. 8.16 Requirement of the truss model.

(c) Model requirement: width (*d*) ≥5 cm, height (*h*) ≥2 cm, length
    (*l*) ≥22 cm (Fig. 8.16);

(d) A loading method and procedure is suggested in Fig. 8.17, which
    can be changed accordingly based on the suggestions from the
    lecturer;

(e) Student evaluation: (i) design report (design idea and consider-
    ations), (ii) question and discussion, and (iii) loading test.

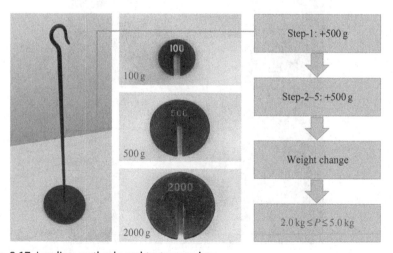

Fig. 8.17 Loading methods and test procedure.

REFERENCES

Brockenbrough, R.L., Merritt, F.S., 1999. Structural Steel Designer's Handbook, third ed. McGraw-Hill, New York. ISBN: 0-07-008782-2.
Tachibana, Y., 2000. In: Nakai, H., Kitada, T. (Eds.), Bridge Engineering. Kyoritsu Shuppan, Tokyo.

CHAPTER NINE

Arch Bridges

9.1 INTRODUCTION

An arch is a curved structure that support the loads parallel to its axis of symmetry, and a bridge with an arch as its load carrying system is called an arch bridge. An arch bridge generally has abutments at each end, and works by transferring the self-weight and other external loads in vertical directions partially into a horizontal thrust restrained by the abutments (or piers) at both sides. In addition to compressive forces in axial direction, the arch usually also needs to resist the bending moments and shear forces. The arch bridges can either be built as one-span with two abutments, or can be made from a series of continuous arches, like the Queen's Bridge at Belfast, as shown in Fig. 9.1.

The arch bridges have obvious advantages in comparison with other structural types: (1) the cross-section of the arch rib is mainly subject to compression. The material properties can be fully used, thus the arch bridges generally have relatively large span capacity; (2) relatively large rigidity and small deformation in service stage, thus better driving conditions in comparison with other bridge types, especially cable-supported bridges; (3) obtaining raw material locally, better durability, and less maintenance; (4) highness.

On the other hand, the arch bridges also have obvious disadvantages, such as (1) relatively heavy deadweight, resulting in large horizontal reactions (this is not the case for arch bridges without thrust forces). Though such forces can be used to reduce the sectional bending, good foundations are required; (2) also due to the horizontal reaction forces, special measures should be taken on the piers to avoid the progressive collapse of continuous arch bridges; and (3) Height of the deck arch bridge is generally relatively high. However, these advantages have been improving, and the arch bridges have been used more and more in the engineering practice.

Arch structures are of different types and are also one of the oldest types of modern bridges, and continue to find new applications in many different fields. Currently, the longest arch bridge in the world is the Chaotianmen

Bridge Engineering
http://dx.doi.org/10.1016/B978-0-12-804432-2.00009-8

Fig. 9.1 The Queen's Bridge at Belfast. *(Photo by Lin.)*

Fig. 9.2 The Chaotianmen Bridge. *(Photo by Yan.)*

Bridge (a road-rail bridge, Fig. 9.2) over the Yangtze River in the city of Chongqing, China, with an effective span of 552 m.

9.2 STRUCTURE FEATURES

The distinguished characteristics of an arch bridge are the presence of horizontal reactions at the ends and relatively small bending moment at any sections. The bending moment caused by the horizontal reactions is

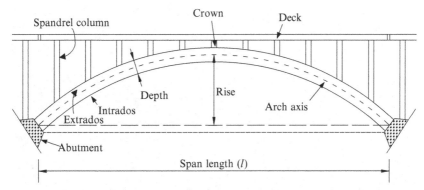

Fig. 9.3 Arch nomenclature.

used to balance the bending moment due to vertical reactions and dead load, live load, etc.

True (or perfect) arch is generally defined as an arch in which only a compressive force acts at the centroid of the arch rib. In a true arch, the dead load produces mainly axial forces, and most of the bending moment is caused by the live load acting over a part of the span. Therefore, there is still bending stress in a true arch, because the real bridge is subjected to different loads including dead load, live load, wind load, and temperature, etc. True arches can be hingeless (fixed) or hinged.

A distinctive terminology originating from the classic masonry arch and relevant terms is shown in Fig. 9.3 (Nettleton, 1977).

9.3 ARCH BRIDGE CLASSIFICATION

The arch bridges may be grouped according to different parameters, such as the shape of the arch rib, namely circular-arc or a parabola-shape. In this section, however, the classification focuses on the construction materials, deck locations, and structural systems.

9.3.1 According to Construction Materials

Arch bridges have been built since ancient times due to easy accessibility of stone masonry, which is an appropriate material for sustaining compressive forces. The Aqueduct Bridge (or the Aqueduct of Segovia) in Spain is a Roman aqueduct and one typical and best-preserved ancient stone arch bridge. The Nihonbashi Bridge (Fig. 3.1), the Ponte Sant'Angelo in Rome (Fig. 9.4), and the Forbidden City Bridge in China (Fig. 9.5) are also typical stone arch bridges. In China, the oldest existing stone arch bridge is the Zhaozhou Bridge of 605 AD. It was designed with perforated spandrels

Fig. 9.4 The Ponte Sant'Angelo, Rome. *(Photo by Yoda.)*

Fig. 9.5 An arch bridge in the Forbidden City, Beijing, China. *(Photo by Yoda.)*

allowing a greater passage for floodwaters. Arch bridges designed in this type can be found worldwide, like the Stone Dock Bridge in China (Fig. 9.6), the Bridge of Arta in Greece, and the Cenarth Bridge in Wales, etc. In 1634, the Spectacle Bridge (a stone arch bridge) was constructed in Japan, as shown in Fig. 9.7. The bridge gets its name from its resemblance to a pair of spectacles when the arches of the bridges are reflected as ovals on the surface of the river. Several stone bridges have been built in Japan following the construction of this bridge.

In addition, the arch bridges can also be built with timber because of its high strength to density ratio, but special attention shall be given to its anisotropic behavior. The Kintai Bridge in Japan is a model timber arch bridge, as shown in Fig. 9.8.

Fig. 9.6 The Stone Dock Bridge, Nanjing, China. *(Photo by Lin.)*

Fig. 9.7 The Meganebashi (or Spectacles Bridge) in Nagasaki, Japan. *(Photo by Yoda.)*

In more modern times, stone and timber arch bridges continued to be built. In addition, other materials like cast iron, steel, and concrete were also increasingly used for the construction of arch bridges. By the end of the 18th century, arch bridges began to be built with iron. The Iron Bridge across the River Severn in England was opened in 1781, which became the first arch bridge in the world made of cast iron, as shown in Fig. 9.9. However, more modern arch bridges are mainly built with reinforced concrete and structural

Fig. 9.8 The Kintai Bridge, Yamaguchi Prefecture, Japan. *(Photo by Lin.)*

Fig. 9.9 The Iron Bridge, Shropshire, UK. *(Photo by Yoda.)*

steel, due to the benefits they give, namely the opportunity for slender, elegant arches, and make longer capacity become possible.

9.3.2 According to Deck Locations

Similar to the classification of truss bridges, the arch bridges can also be classified according to the position of the arch relative to the deck, including deck arch, through arch, and half-through arch bridges as shown in Fig. 9.10.

The deck arch bridge represents an arch bridge in which the deck is completely above the arch. The components between the arch and the deck are named as the spandrel. The spandrel can be solid (mainly used for stone or masonry arch bridges), and usually called as the closed–spandrel deck arch bridge. If the spandrel is made of a number of vertical columns, then it is called as an open spandrel deck arch bridge. For the case that an arch only supports the deck at the crown (peak of the arch), it is named as a cathedral arch bridge.

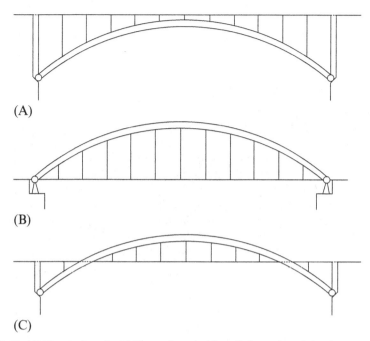

(A)

(B)

(C)

Fig. 9.10 (A) The deck arch. (B) Through arch. (C) Half-through arch bridges.

For an arch bridge in which the deck is at the arch base and passes through the arch, it is often called a through arch bridge. In this case, the deck is generally supported by suspension cables or tie bars. If the bridge deck is placed above the arch base but below the crown (the deck still passes through the arch), it is often referred to as a half-through arch bridge. Some examples are shown in Figs. 9.11 and 9.12.

9.3.3 According to Structural Systems

An arch system can be grouped as a fixed arch (Fig. 9.13A), two-hinged arch (Fig. 9.13B), or three-hinged arch (Fig. 9.13C) according to the number of hinges.

The fixed arch (or hingeless arch) is fixed at the abutments so that moment is transmitted to the abutment. The fixed arch has three redundancies, allows no rotation at the foundations. Fixed arch is a very stiff structure and suffers less deflection than other arches. However, as fixed arch is a structurally indeterminate structure, a great deal of forces will be generated at the foundation. Therefore, fixed arch bridges can only be built where the

Fig. 9.11 Deck arch and half-through arch (the Komagata-bashi Bridge in Tokyo). *(Photo by Lin.)*

Fig. 9.12 A through arch bridge (the Kachidoki-bashi Bridge in Tokyo). *(Photo by Lin.)*

ground is very stable. The fixed arch is most often used in reinforced concrete bridges, where the spans are short.

The hinged arches involve three hinge arrangements: single-hinged type, two-hinged type, and three-hinged type (Xanthakos, 1993). In arch bridges, two hinges or three hinges are frequently used. The hinge used in a steel arch bridge is shown in Fig. 9.14. The two-hinged arch has pins at the end bearings, so that only horizontal and vertical components of force act on the abutment. The two-hinged arch is most often used to bridge long spans. The three-hinged arch has a hinge at the crown as well as the abutments, making it statically determinate and eliminating stresses due to change of temperature and rib shortening. In addition, the less complex forces on the bases can simplify the foundation design. Three-hinged arch also has obvious drawbacks. For example, three-hinged arch bridges have smaller rigidities and therefore experience much more deflection. In

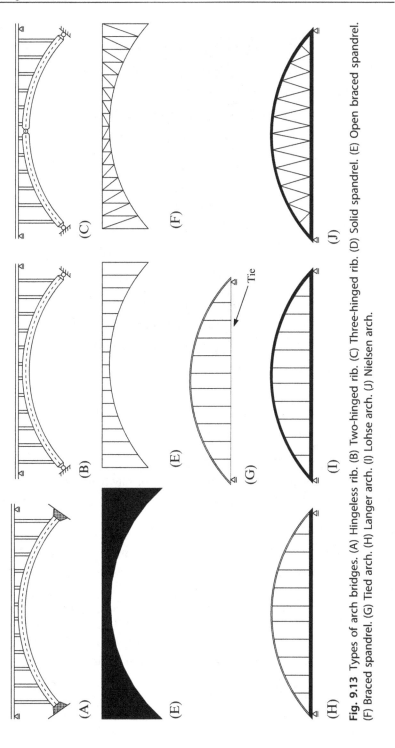

Fig. 9.13 Types of arch bridges. (A) Hingeless rib. (B) Two-hinged rib. (C) Three-hinged rib. (D) Solid spandrel. (E) Open braced spandrel. (F) Braced spandrel. (G) Tied arch. (H) Langer arch. (I) Lohse arch. (J) Nielsen arch.

Fig. 9.14 Hinges in an arch bridge. *(Photo by Lin.)*

addition, the hinges are complex in fabrication. Steel arch bridge can be built as either hingeless (fixed) or hinged.

Arch bridges can be built with solid spandrel, open spandrel, and braced spandrel (Fig. 9.15). Solid spandrel is generally built with two hinges or is hingeless, and the masonry arch bridges are generally built with solid spandrel. The spandrel can be built as open with vertical columns, referred to as open spandrel. The spandrel can also be designed in the shape of a truss with curved lower chords. This type of arch is generally constructed with two or three hinges.

In traditional ideas, the arch bridges have horizontal thrust force under loads in vertical direction. The arch bridges can also be "stiffened" and such bridges are referred to as stiffened arch bridges, like tied-arch, Langer arch, Lohse arch, and Nielsen arch bridges, etc. A tied-arch bridge is an arch bridge in which the horizontal forces are resisted by tie-rods, rather than by the bridge foundations, as shown in Fig. 9.13G. The elimination of horizontal forces at the abutments allows tied-arch bridges to be constructed with less robust foundations. In addition, since they do not depend on horizontal compression forces for their integrity, tied-arch bridges can be prefabricated off-site, and subsequently transported to the construction site.

Fig. 9.15 An arch bridge with braced spandrel. *(Photo by Lin.)*

In a Langer arch bridge, the arch rib is thin but the girders are deep. Therefore, it is assumed that the arch rib only resists axial compression, while the girders take both bending moment and axial tension. In a Lohse arch bridge, however, it is generally assumed that both the arch rib and girders will take both bending moment and axial forces. The arch shown in Fig. 9.13J is referred to as a Nielsen arch. The deck arch bridges with Langer and Lohse archers are called reversed Langer arch and reversed Lohse arch bridges, respectively.

Some stiffened arch bridges are shown in Fig. 9.16. In addition to vertical arch ribs, there are also arch bridges having inclined ribs, such as the bridges in Fig. 9.17.

9.4 ERECTION OF ARCH BRIDGES

For an arch bridge, both its greatest advantage and biggest inconvenience are due to its shape. An arch shape has obvious advantages, but it has to be completed in order to be functional. A partial arch during construction of the arch has little to do with the final structure, and the construction method is always a concern when an arch is selected.

A similar construction method may be used in both steel and concrete arch bridges, though they were generally first used in steel bridges and then later applied to the concrete ones. This is due to the fact that steel arch bridges have been used much earlier than concrete arch bridges. Also, steel arch has lighter weight than that of concrete arches. In general, the arch bridge construction can be classified according to the use of falsework. The stone

Fig. 9.16 Arch bridge examples. *(Photos by Taniguchi, Yoda, and Lin.)*

Fig. 9.17 Arch bridges with inclined ribs. *(Photos by Taniguchi and He.)*

arch bridges and concrete arch bridges built with precast segments are usually constructed on falsework, while other arch bridges are constructed without falsework.

9.4.1 Arch Construction With Falsework

Dating back to the early arch bridges, the most used method of building masonry or concrete arches has been on timber falsework. This is a system easy to use that is still used for the construction of small and medium span arches. However, with the increase of the arch span, the construction cost increases rapidly due to excessive consumption of the falsework.

In setting the elevation of the soffit forms, the deformation of the timber supports must be taken into account. The actual movement during masonry materials or arch segment placement should be monitored and adjustments made by the jacks if needed (Nettleton, 1977).

9.4.2 Arch Construction Without Falsework

There are many methods for arch bridge construction without need of falseworks, and the following are given as examples.

9.4.2.1 Free Cantilevers Construction

This method is mainly used for steel arches made of lattice structures. The cantilevers of the semiarches do not require any falsework, thus a construction in free cantilevers becomes possible. Typical arch bridges built by this method include the Sydney Harbour Bridge built in 1932 (Fig. 9.18). The semiarches were provisionally built in lengthening the arch's upper

Fig. 9.18 The Sydney Harbour Bridge. *(Photo by Taniguchi.)*

chord by means of stays, and the semiarches were then completed using the free-cantilevered construction system.

9.4.2.2 Cantilever Tieback Construction
This method is widely used in the construction of large arch bridges. Temporary towers are built on the abutment or piers to support steel tie bars. It consists of provisionally cable-staying the subsequent cantilevers of the semiarches until closing them at the keystone.

The Eads Bridge over the Mississippi in St. Louis started in 1868, was the first to use this method of erection. This bridge was designed as three arches of $152 + 157 + 152$ m spans. Since this is a multiple span bridge, balanced cantilever erection was used for the intermediate arches. In addition, this is also the first bridge to be built using free cantilevers. After this bridge was built, the free-cantilevered construction procedure quickly spread to all bridge types. Since then, many long span arch bridges have been erected by the cantilever tieback method, such as the Rainbow Bridge completed in 1941 and the Glen Canyon Bridge in 1958, and the New River Gorge Bridge built in 1976, etc.

Stability of the arch bridge during construction should be carefully calculated, and transverse connections are generally used between two arch ribs.

9.4.2.3 Off-Site Construction
This method is mainly used for self-equilibrated arch system, such as tied arches that require only vertical forces. Such arch bridges can be fabricated at a factory and transported to the bridge construction site. After that, they are lifted and placed at their final position.

9.4.2.4 Rotation of the Semiarches
This procedure consists of building two semiarches in a quasi-vertical position over the abutments and rotating them later on by means of a back stay, until closing them with a keystone. This method can be further divided into plane rotating method, vertical rotating method, and combined plane and vertical rotating method. The Wushan Longmen Bridge built in 1986 in China is a typical arch bridge built by using this method.

The Dashengguan Yangtze River Bridge, the Guangyuan Jialing River Bridge, and Zhaohua Jialing River bridge during construction are shown in Figs. 9.19–9.21.

Fig. 9.19 The Dashengguan Yangtze River Bridge. *(Photo by He.)*

Fig. 9.20 The Guangyuan Jialing River Bridge. *(Photo by Gui.)*

Fig. 9.21 The Zhaohua Jialing River Bridge. *(Photo by Gui.)*

9.5 CASE STUDY: PRESERVATION OF MASONRY ARCH BRIDGES

The Nihonbashi Bridge is the point from which all distances are measured to the capital; highway signs indicating the distance to Tokyo actually state the number of kilometers to Nihonbashi. Many people recognize the need to preserve the historic stone arch bridges, as they are part of the history and heritage of many local communities. Although stone is one of the oldest construction materials, the preservation and rehabilitation of historic stone arches requires professional skill as well as empirical knowledge.

From the view point of bridge engineering, understanding the appropriate evaluation methods, analysis techniques, and repair and strengthening method is very important for the preservation and rehabilitation of stone masonry arch bridges.

Structural assessment of stone masonry arch bridges is the field investigation of the bridges. Requires properties to be investigated are the overall geometry, thickness of the arch barrel, and materials used in the construction.

Stone arch bridges can be analyzed using different analysis techniques. The most common techniques are the finite element (FE) or discrete element (DE) methods. The finite element method is suitable for all structures, including problems involving complex geometry or load conditions and evaluation of various strengthening options.

The evaluation technique requires the load-carrying capacity to be evaluated using the allowable stress (AS) method based on limiting the tensile and compressive stresses that develop in extreme fibers under a combination of axial and bending forces (Figs. 9.22 and 9.23).

With a view to numerical results, repairing involves localized mortar repointing and replacement of deteriorated stone. Often the new stone can be sourced locally and brought in and shaped as necessary to achieve an appearance matching the original stone. A properly functioning drainage plan is important to mitigate the risk of material deterioration in the arch (Fig. 9.24).

The structural capacity of an existing arch bridge can be improved in several ways. A new structural slab can be built with waterproof sheet (Nagata, 2011), as shown in Fig. 9.25.

(A)

(B)

Fig. 9.22 Masonry stone arch bridge (Nihonbashi Bridge). (A) After cleaning. (B) Before cleaning.

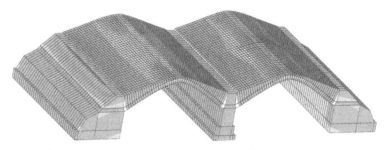

Fig. 9.23 Stress distribution in the vaults and spandrel walls.

Fig. 9.24 Water leakage from the above before repairing.

Fig. 9.25 Waterproof sheet.

9.6 EXERCISES

1. What is a true arch? Give some true arch examples in the world.
2. Determine the bending moment (M) and axial force (N) influence lines in a three hinge arch.

REFERENCES

Nagata, Y., 2011. Preventive maintenance work of Nihonbashi Bridge. Bridge Found. Eng 45 (4), 42–46 (in Japanese).

Nettleton, D.A., 1977. Arch Bridges. Bridge Division, Office of Engineering, Federal Highway Administration, U.S. Dept. of Transportation, Washington, DC.

Xanthakos, P.P., 1993. Theory and Design of Bridges, first ed. Wiley-Interscience, New York.

CHAPTER TEN

Cable-Stayed Bridges

10.1 INTRODUCTION

A cable-stayed bridge is a structural system with a continuous girder (or bridge deck) supported by inclined stay cables from the towers (or pylons), as shown in Fig. 10.1. Form the mechanical point of view, the cable-stayed bridge is a continuous girder bridge supported by elastic supports. The cable-stayed bridge ranks first for a span range approximately from 150 to 600 m, which has longer spanning capacity than that of cantilever bridges, truss bridges, arch bridges, and box girder bridges, but shorter than that of suspension bridges. In this range, the cable-stayed bridge is very economical and has elegant appearance due to the relatively small girder depth and has proved to be very competitive against other bridge types. In addition, with the development in bridge design and construction, more and more cable-stayed bridges are being built with longer spans. Currently, the Russky Bridge in Russia is the largest cable-stayed bridge with the world's longest span, at 1104 m.

The concept of a cable-stayed bridge is simple, as all the members in a cable-stayed bridge mainly work in either tension or compression. The stay cables provide intermediate elastic support for carrying the vertical loads acting on the main girder so that it can span a longer distance. To carry the loads applied on the bridge deck, the cables need to sustain the tensile axial force, which therefore results in compression forces in both pylons and main girders. Though there are also some bending moments or other forces in pylons and the main girders, generally their effects are much smaller than that of the axial forces. It is well known that axially loaded members are more efficient than flexural members, which contributes to the structural efficiency and economy of a cable-stayed bridge.

The earliest design of cable-stayed bridges dates back to 1595, which is evident from the book of Machinae Novae. Several cable-stayed bridges were built in the early part of 19th century, but it was not until the 1950s did the start becoming prevalent like other bridge types, such as truss

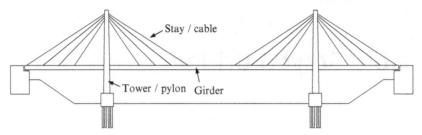

Fig. 10.1 A typical cable-stayed bridge.

bridges, arch bridges, and suspension bridges. Several cable-stayed bridges collapsed due to lack of understanding of such a system, particularly due to inadequate resistance since it was not possible to tension the stays, and they would become slack under various load conditions. These ideas were not well adapted and improved until the construction of the Brooklyn Bridge, completed in 1883 (ICE, 2008).

The modern concept of the cable-stayed bridge was first proposed in Germany in the early 1950s, and the first modern cable-stayed bridge was the Stromsmund Bridge in Sweden built in 1955. But thereafter, both design and construction of cable-stayed bridge have developed much faster than any other bridge types during this period. The cable-stayed bridges have been used worldwide nowadays.

10.2 CABLE-STAYED BRIDGE CLASSIFICATION

10.2.1 Stay Cable Arrangements

According to the longitudinal cable layout, the cable-stayed bridges can be classified into four types: mono, fan, modified fan, and harp, as shown in Fig. 10.2. All these cable configurations have been used in practice, but the cable configuration generally does not have a major effect on the behavior of the bridge except in very long span structures (Tang, 1999).

10.2.1.1 Mono Cable System

The mono design uses a single cable from its towers, and is one of the lesser-used cable-stayed bridge type and is rarely built. The Neckar River Bridge in Germany is a typical mono cable-stayed bridge. Cable-stayed bridges with less number of cable stays are occasionally used, like the Shin-Ohashi Bridge in Tokyo and the Toyosato-Ohashi Bridge in Osaka, as shown in Figs. 10.3 and 10.4. Early cable-stayed bridges are designed with very few stay cables,

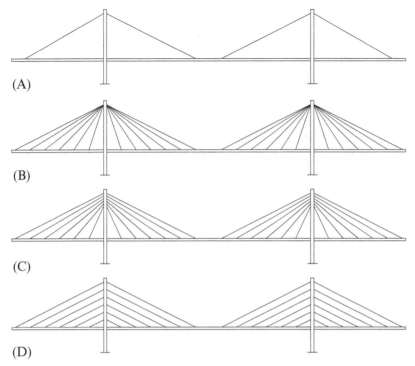

Fig. 10.2 Longitudinal cable arrangements. (A) Mono. (B) Fan (radial). (C) Modified fan. (D) Harp.

Fig. 10.3 The Shin-Ohashi Bridge in Tokyo. *(Photo by Lin.)*

Fig. 10.4 The Toyosato-Ohashi Bridge in Osaka. *(Photo by Lin.)*

but this involves substantial erection costs. Modern cable-stayed bridges tend to use many more cables to ensure greater economy.

10.2.1.2 Fan Cable System

In the fan design, all stay cables connect to or pass over the top of the towers. The fan design is structurally superior with minimum moment applied to the towers. In addition, due to the steeper cable slopes, the fan design is structurally efficient with maximum vertical component for sustaining the vertical loads, but smallest axial force applied to the main girders. The fan system was adopted for several of the early designs of the modern cable-stay bridges, including the Stromsmund Bridge (Wenk, 1954; ICE, 2008). However, there are also obvious difficulties in practical application due to the possible corrosion and fatigue problems at the pylon head caused by the large amount of cable forces. In addition, the anchorages are generally very heavy and complicated and the tower is needed to be further strengthened at the termination point.

10.2.1.3 Modified Fan Cable System

To avoid the difficulties in a fan cable system due to the "fixation together" of cable stays, the modified fan cable system is developed. In this system, the cables connect near the top of the tower but is spaced sufficiently from each other for the benefits of better fixation like improved force transmission, and easy access to individual cables for inspection and maintenance. The modified fan is particularly suitable for bridges with many cables. The modified fan cable system was used in the Ting Kau Bridge, as shown in Fig. 10.5.

10.2.1.4 Harp Cable System

In the harp cable system, the cables are nearly parallel to each other so that the height of their fixed points on the tower is proportional to the distance

Fig. 10.5 The Ting Kau Bridge in Hong Kong (1998). *(Photo by Yoda.)*

from the tower to their positions on the deck. A harp cable system offers a very neat and orderly arrangement with elegant appearance because an array of parallel cables will always appear parallel irrespective of the viewing angle. It also allows an earlier start of girder construction because the cable anchorages in the tower begin at a lower elevation (Tang, 1999). The Jiuzhou Port Channel Bridge in Hong Kong-Zhuhai-Macao Bridge was designed with the Harp cable system, as shown in Fig. 10.6.

10.2.2 Lateral Cable Arrangements

In the lateral direction, the cable system can be arranged as one single plane above the center line, two planes (either vertical or inclined) at the edges of the girder, or three planes connect the centerline and both edges of the girder. The layout of the cable stays affects the structural behavior, the erection method, and the architectural expression of the bridge. The most common solution is to construct the bridge with two cable planes, but some bridges have been built with one central plane. The major problem of a central cable plane system is its insufficient torsional stiffness to resist the twisting moment from eccentric loading, e.g., traffic load in only one side of the carriageway.

Fig. 10.6 The Jiuzhou Port Channel Bridge in Hong Kong-Zhuhai-Macao Bridge. *(Photo by Lin.)*

If it is the case, a box-section may be used to achieve the required torsional stiffness. For two plane cable system, however, this is not necessary because the cable system can provide necessary torsional stiffness for cable-stayed bridges. In case the deck of the bridge is very wide it is possible to design three cable planes because transverse bending moment is reduced when the deck is divided into two parts with three cable planes, as shown in Fig. 10.7. If the bridge is for both railway and road, the railway can be placed in the middle of the deck between the cable planes while the lanes are on cantilever in lateral direction.

10.2.3 Number of Spans (Or Towers)

The cable-stayed bridge can be designed as single span, two spans, three spans, or multiple spans. However, cable-stayed bridges having either three or two cable-stayed spans are more widely used, which is because the cable stays and the anchor pier are important for the stability of the pylon. The cable-stayed bridges can be built with only one tower, like the Erasmus Bridge in Rotterdam, the Toyosato-Ohashi Bridge in Osaka, and the Chuoohashi Bridge in Tokyo, as shown in Figs. 10.8–10.10. When a bridge has more than three spans, the main problem is the lack of longitudinal restraint to the top of the intermediate pylons, which cannot be directly anchored to an approach pier. Large deformations can occur in multiplespan cable-stayed bridges under the live load. As shown in Fig. 10.11, this problem can be solved in the following ways: (a) by increasing the stiffness of

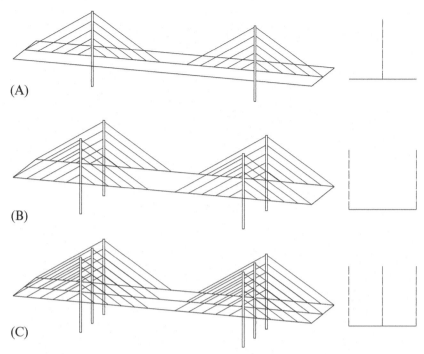

Fig. 10.7 Lateral cable arrangements. (A) One center plane (singe plane). (B) Two lateral planes (double planes). (C) Three planes (triple planes).

Fig. 10.8 The Erasmus Bridge in Rotterdam (1998). *(Photo by An.)*

Fig. 10.9 The Tsuneyoshi Ohashi Bridge in Osaka. *(Photo by Lin.)*

Fig. 10.10 The Chuoohashi Bridge in Tokyo. *(Photo by Lin.)*

pylons. As the increase of the deck may generally cause the increase of the girder section and thus the dead load, it would be better to stiffen the pylon. For example, by using the A frame braced pylon as shown in Fig. 10.11A; (b) by using additional horizontal cables between tower tops, directly transfer any out-of balance forces to the anchor stays in the end spans; (c) by using additional cables to connect the top of the internal pylons to the adjacent pylon at deck level so that any out-of-balance forces are resisted by the stiffness of the pylon below deck level; (d) by using additional tie-down piers at span centers; or (e) by adding additional cables at the midspans (Tang, 1995). These cables cross each other and extend for approximately 20% of span length beyond the span center. The advantage of such a method is that it is efficient in reducing the bending moments in the girder and the towers to an acceptable level and still retains the slender look of a conventional cable-stayed bridge. The Ting Kau Bridge in Hong Kong is such a typical example. In order to stabilize the central tower, longitudinal stabilizing

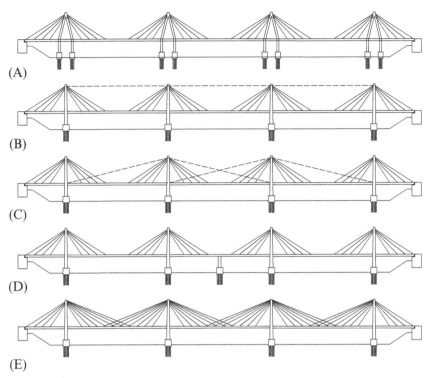

(A)

(B)

(C)

(D)

(E)

Fig. 10.11 Multiple span cable-stayed bridges. (A) A-frame braced pylon. (B) Additional cable system between pylon tops. (C) Additional cable system between intermediate pylon and girders. (D) Additional tie-down piers at span centers. (E) Additional cables at the mid-spans.

Fig. 10.12 The Ting Kau Bridge in Hong Kong. *(Photo by Lin.)*

cables up to 464.6 m long have been used to diagonally connect the top of the central tower to the deck adjacent to the side towers, as shown in Fig. 10.12.

10.3 CONFIGURATION

10.3.1 Cable

The cable stays are the key load carrying and transferring members in cable-stayed bridges, and the main problems with early cable-stay bridges were deficiencies with the anchorage system, steel material, and corrosion. Available stay systems for modern cable-stayed bridges include prefabricated locked coil (modern locked coil stays provide all the wires in a finally galvanized condition and will achieve a tensile strength of up to 1770 N/mm^2), prefabricated helical, or spiral strand by using wire of 5 mm diameter and a tensile strength of either 1570 N/mm^2 or 1770 N/mm^2, bar bundles (steel bars with a tensile strength of 1230 N/mm^2), parallel wire strand (PWS, most commonly comprises 7 mm diameter galvanized round steel wires with a tensile strength of 1570 N/mm^2), parallel strand (usually manufactured from 15.2 or 15.7 mm diameter galvanized seven-wire strands with a tensile strength of 1770 N/mm^2), and advanced composite stays, etc. (ICE, 2008). Typical stay types used for cable-stayed bridges are shown in Fig. 10.13.

10.3.2 Pylon

The pylons can be designed as a single column projecting through the center of the deck, but sometimes located on one side, such as in curved cable-stayed bridges. It is also possible to arrange a pair of columns (without lateral beams

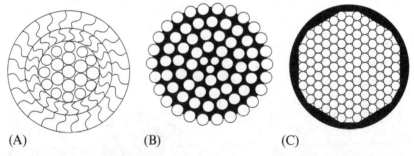

Fig. 10.13 Stay types for cable-stayed bridges. (A) Locked coil cable. (B) Spiral strand cable. (C) Parallel wire strand (PWS).

between them) on both sides of the deck. The H-shape pylon is similar to the double arrangement, but has a lateral member connecting the two columns. The A-shaped design is similar in concept to the H-shape pylon, but with two inclined columns toward each other and meeting at the top, eliminating the need for the third member. Depending on the design, the columns may be angled or vertical below the bridge deck. To be specific, bridge towers can be designed as H frame, A frame, and inverted Y frame pylons, diamond pylon, and twin-diamond pylon, as shown in Fig. 10.14. Some cable-stayed bridges with different pylons are shown in Figs. 10.15–10.19.

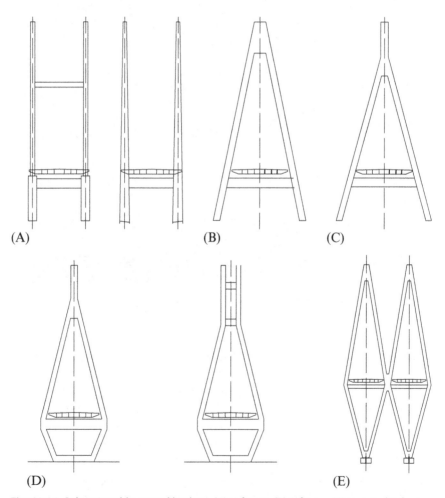

Fig. 10.14 Pylons in cable-stayed bridges. (A) H-frame. (B) A-frame. (C) Inverted Y-frame pylons. (D) Diamond pylon. (E) Double diamond pylon.

Fig. 10.15 The Sugaharashirokita Oohashi Bridge in Osaka. *(Photo by Lin.)*

Fig. 10.16 The Yanba Bridge, Gunma Prefecture. *(Photo by Taniguchi.)*

Early cable-stayed pylon designs were predominantly constructed as steel boxes with thick steel plates joined together by either welding or riveting. The advantage of metal pylons lies in their quicker fabrication and erection, but the buckling of the tower is a concern. Pylons, especially large cable-stayed bridges, can also be built more economically with reinforced or prestressed concrete. Thus the pylons can be built with concrete than with steel in order to reduce the cost. Advances in concrete construction and modern formwork technology have made the use of concrete increasingly competitive for construction of pylon with complex forms. Though the concrete tower has larger self-weight, they are strong for sustaining

Fig. 10.17 The Second Severn Crossing. *(Photo by Yoda.)*

Fig. 10.18 The Qingzhou Channel Bridge in Hong Kong-Zhuhai-Macao Bridge. *(Photo by Gui.)*

the load in compression. The pylon used in Jiuzhou Port Channel Bridge can be found in Fig. 10.20.

The pylon height is generally taken as 0.2–0.25 of the main span length, and the efficiency of the stay is not significantly impaired when the stay inclination is varied within reasonable limits, which may be taken as 25–65 degrees. The height of pylon may also be affected by other factors. For example, the cable-stayed bridge to be constructed at Kawasaki was designed with low towers because its location is close to the Haneda international airport.

Fig. 10.19 The Bangkok Ring Road Cable-stayed Bridge, Thailand. *(Photo by Yoda.)*

Fig. 10.20 The Jiuzhou Port Channel Bridge in Hong Kong-Zhuhai-Macao Bridge. *(Photo by Lin.)*

10.3.3 Deck

In general, the deck needs to resist both bending moment from the dead-weight and live load, and axial force derived from the horizontal component of the stay force. Therefore, unlike the deck in a suspension bridge, the deck can be designed as different sections (or structural forms) in cable-stayed bridges. The often used deck sections include the steel deck, the composite deck, and the concrete deck. The deck will influence the whole structure of a cable-stayed bridge due to its characteristics of self-weight and aerodynamics.

10.3.3.1 Steel Deck

For early cable-stayed bridges, the steel deck was used due to its high load-capacity to weight ratio and larger span capacity between cable stays. In addition, the reduction in deck weight can result in an economic design for large span bridges, as in the Tatara Bridge. The orthotropic road deck generally consists of a thin surfacing material laid on steel plate stiffened longitudinally. The stiffeners are supported by transverse floor beams. The design of the deck cross section is dominated by the arrangement of the stays. The steel deck used for the Jianghai Channel Bridge in Hong Kong-Zhuhai-Macao Bridge is shown in Fig. 10.21.

10.3.3.2 Concrete Deck

Reinforced or prestressed concrete decks can be made of precast elements or they can be cast in place. The concrete deck is suitable for medium spans because the cost of concrete is relatively low but its weight increases the dead

Fig. 10.21 The Jianghai Channel Bridge in Hong Kong-Zhuhai-Macao Bridge. *(Photo by Lin.)*

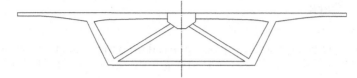

Fig. 10.22 Concrete torsion box deck.

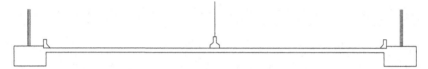

Fig. 10.23 Twin beam concrete deck (ICE, 2008).

load of the bridge thus requiring larger dimensions for cables, pylons, piers, and anchorage structures. For a cable-stayed bridge with single central plane of stays, a strong torsion box-section is needed to provide the torsional resistance, as shown in Fig. 10.22. While for a cable-stayed bridge with multicable system, the deck will be an open girder cross section bridges with very long spans should use cross sections with high torsional stiffness. The twin beam concrete deck used in Dames Point Bridge over St Johns River in Florida, USA, provided as an example of a simplified deck form, as shown in Fig. 10.23.

10.3.3.3 Composite Deck

The deck in cable-stayed bridges can also be built as steel-concrete composite section. Composite construction of steel and concrete is a popular structural method due to its numerous advantages against conventional solutions. The optimal combination of the properties of the two most popular construction materials, i.e., steel and concrete, results in structures that are both safe and economic (Vasdravellis et al., 2012). In cable-stayed bridges, the composite concrete slab over the steel orthotropic deck provides a new option. In composite bridges the anchors can be aligned with the stiffening girder or placed in an exterior position (under or in the slab plan).

To minimize the displacement in the middle span, a combination of deck types such as steel deck, concrete deck, and composite deck can be used for the mid-span and side spans. In such a case, heavier section (i.e., concrete section or composite section) should be used in side span, while lighter section (i.e., steel section or composite section) should be used in midspan to

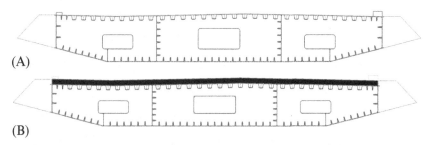

Fig. 10.24 Steel deck and composite deck. (A) Steel section for middle span. (B) Composite section for side span.

reduce the down-ward deflection in midspans and avoid the upward defection in side spans. The proposed sections for a cable-stayed bridge are shown in Fig. 10.24, in which the steel section was proposed for the midspan to reduce the self-weight, and the composite section was designed for the side span.

10.4 ANALYSIS OF CABLE-STAYED BRIDGES

Both static and seismic analyses should be performed on cable-stayed bridges. For the analyses of modern cable-stayed bridges, finite element analyses are always necessary. The pylon, deck, and the cable stays will be modeled by suitable element, and the "fish-bone" model is usually used to simulate the whole bridge. The stays can be represented with a small inertia and a modified modulus of elasticity that will model the sag behavior of the stay. In addition, for considering the force transformation and load redistribution during the erection, stage-by-stage phase analyses is always necessary. A typical "fish-bone" model of a cable-stayed bridge is illustrated in Fig. 10.25, and both linear and nonlinear analyses can be performed by using FEM software packages. In addition to the static analyses, the dynamic analyses for determining the dynamic performance of cable-stayed bridge, such as frequencies and the vibration modes, should also be performed.

10.5 CONSTRUCTION OF CABLE-STAYED BRIDGES

The success of the cable-stayed bridge is due to its efficient erection procedure. Depending on whether the temporary supports are used or not, there are two construction methods: erection with temporary supports or free cantilever method (Gimsing and Georgakis, 2012).

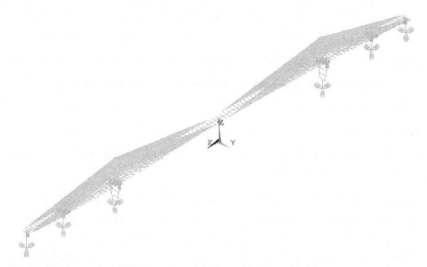

Fig. 10.25 The typical "fish-bone" model of a cable-stayed bridge.

10.5.1 Erection With Temporary Supports

In this method, temporary supports are used for deck erection before adding and tensioning the cable stays. The whole construction can be conducted in the following stages: (1) Erection of the deck on the permanent piers and the temporary supports; (2) Erection of the pylons from the completed deck; (3) Installation and tension of the stay cables; and (4) dismantlement of the temporary supports. This method is simple and easy to use, but temporary supports that are required and clearance requirements for navigation during the construction should be considered.

10.5.2 Free Cantilever Method

Free cantilever method is more frequently used for modern cable-stayed bridge construction, in which bridge deck is directly supported by the cable during construction. In this case, the cable-stayed bridge is always in a cantilever condition before the deck erection is completed. The construction can be carried out in the following stages: (1) pylons and the deck units above the main piers are erected and fixed to the piers; (2) new deck segments are erected by free cantilevering from the pylon, either symmetrically in both directions or only into the main span. Simultaneously, the stay cables are installed and tensioned initially to relieve the bending moments in the deck; and (3) the stage-2 is repeated until the deck at midspan are connected, as shown in Fig. 10.26. For this method, the construction safety, especially of the final stage before the connection of the deck at midspan (largest cantilever condition), should be carefully confirmed.

Some pictures of cable-stayed bridges during construction are shown in Fig. 10.27.

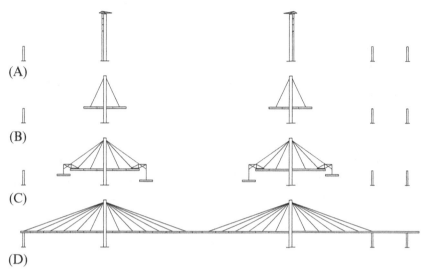

Fig. 10.26 Construction of a cable-stayed bridge by free cantilever method. (A) Stage-1. (B) Stage-2. (C) Stage-3. (D) Stage-4.

Fig. 10.27 Pictures of cable-stayed bridges during construction.

10.6 EXERCISES

1. Describe the structural components of a typical cable-stayed bridge and the internal forces they need to sustain.
2. Describe possible lateral and longitudinal layouts of cables in the cable-stayed bridges.

REFERENCES

Gimsing, N.J., Georgakis, C.T., 2012. Cable Supported Bridges: Concept and Design. Wiley, Chichester.
Institution of Civil Engineers, 2008. ICE Manual of Bridge Engineering, second ed. Thomas Telford, London.
Tang, M.C., 1995. Multispan cable-stayed bridges. In: International Bridge Conference Bridges into the 21st Century, Hong Kong, Oct. 1995.
Tang, M.C., 1999. Chapter 19: cable-stayed bridges. In: Chen, W.F., Duan, L. (Eds.), Bridge Engineering Handbook. CRC Press, Boca Raton, FL.
Vasdravellis, G., Uy, B., Tan, E.L., Kirkland, B., 2012. Behaviour and design of composite beams subjected to negative bending and compression. J. Constr. Steel Res. 79, 34–47.
Wenk, H., 1954. The Stromsmund Bridge. Stahlbau 23 (4), 73–76.

CHAPTER ELEVEN

Suspension Bridges

11.1 INTRODUCTION

A suspension bridge is a type of bridge in which the deck is hung by suspension cables on vertical suspenders. The basic structural components of a suspension bridge system include stiffening girders/trusses, the main suspension cables, main towers, and the anchorages for the cables at each end of the bridge. The main cables are suspended between towers and are finally connected to the anchorage or the bridge itself, and vertical suspenders carry the weight of the deck and the traffic load on it. Like other cable supported bridges, the superstructure of suspension bridges is constructed without false work as the cable erection method is used. The main load carrying member is the main cables, which are tension members made of high-strength steel. The whole cross-section of the main cable is highly efficient in carrying the loads and buckling is not problem. Therefore, the deadweight of the bridge structure can be greatly reduced and longer span becomes possible. In addition, the esthetic appearance of suspension bridges is another advantage in comparison with other types of bridges.

The early modern bridges pertaining to this type of bridge were built in the early 19th century (Waddell, 1905; Gerner, 2007). The design drawing of the Chakzam Bridge, south of Lhasa is shown in Fig. 11.1. This bridge has cables suspended between two towers, and vertical suspender cables carrying the weight of a planked footway below. This bridge is quite similar to modern bridges though sections were very simple.

Some suspension bridges were built in early part of 19th century, including the Jacob's Creek Bridge (iron chain suspension bridge built in 1801) in Pennsylvania. Thereafter, the Dryburgh Abbey Bridge (1817), the Union Bridge (1820), and Menai Bridge (1826) were built successively in the UK. The Clifton Suspension Bridge (construction began in 1836 but was interrupted in 1843 due to lack of funds, and finally opened in 1864) is another important (parabolic arc chain type) suspension bridge, as shown in Fig. 11.2.

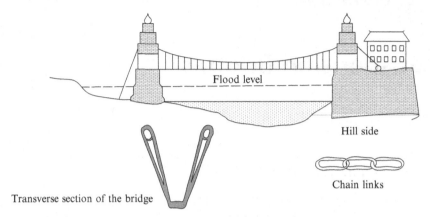

Fig. 11.1 Iron suspension bridge over Lhasa (Waddell, 1905).

Fig. 11.2 The Clifton Suspension Bridge. *(Photo by An.)*

11.2 STRUCTURAL COMPONENTS

The structural components of a modern suspension bridge include the stiffening girder, the main cable, the main tower, the anchorage, and hanger ropes, etc. The stiffening girder is a deck structure together with a longitudinal stiffening system, which is the longitudinal member that supports and distributes vertical live load. The stiffening girder can be either a separate truss or plate stiffening girders combined with lateral bracing systems, or alternatively be integrated with the deck structure in the form of a shallow box girder with a low drag shape to minimize wind loading (ICE, 2008).

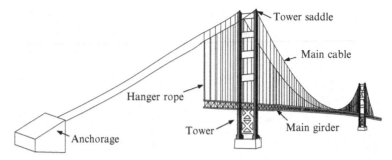

Fig. 11.3 Suspension bridge components.

The main cable made of high-strength steel wires supports traffic-carrying stiffening girder by hanger ropes and transfer loads by direct tension forces to towers and anchorages. A main tower is the intermediate vertical structures, which supports main cables at a level determined considering the cable sag and required clearance, and transfers the external loads to bridge foundations. An anchorage is generally a massive concrete block, which anchor main cables and act as end supports of a bridge against movement in the horizontal direction. The suspension cables must be anchored at each end of the bridge or sometimes to the bridge itself, because all loads applied to the bridge are transformed into a tension in these main cables. The vertical cable (or hanger rope) is the cable connecting the stiffening girder with the main cable, which is mainly used to transfer the live load applied on the deck to the main cable. The structural components of a typical suspension bridge are shown in Fig. 11.3.

In addition to modern highway or railway bridges, suspension bridges are also used as pedestrian bridges, like the bridge shown in Fig. 11.4.

11.3 SUSPENSION BRIDGE CLASSIFICATION

In general, the suspension bridges can be classified according to their span numbers, the connection between stiffener girders, the layout of suspenders, and anchoring conditions, etc.

11.3.1 According to Span Numbers

Based on the number of spans and towers, there are single-span, two-span, or three-span suspension bridges, as shown in Fig. 11.5. Among them, three-span suspension bridges with two main towers are the most commonly used in engineering practice, like the Rainbow Bridge in Tokyo as shown in

Fig. 11.4 A suspension bridge in Izu, Japan. *(Photo by Lin.)*

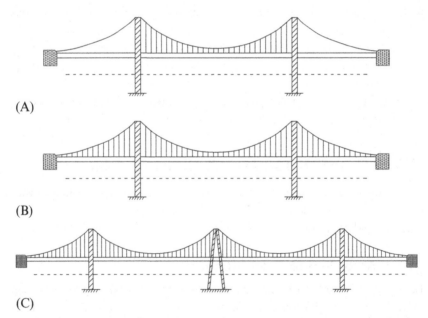

Fig. 11.5 Suspension bridge classification according to span numbers. (A) Single span. (B) Three-span. (C) Four (or multi) span.

Fig. 11.6. The Tsing Ma Bridge in Hong Kong and the Pingsheng Bridge in Guangdong are typical single-span suspension bridges, as shown in Figs. 11.7 and 11.8. For multispan suspension bridges with more than two towers, the horizontal displacement of the tower tops due to live loads can be a concern and measures for controlling such displacement becomes necessary. The Tamate Bridge built in 1928 in Japan is a typical multispan suspension bridge, which is still in use now. Since then, several bridges were built in France (Pont de Châteauneuf-sur-Loire, 1932; Chatillon Bridge, 1951; and Bonny-sur-Loire Bridge, etc.), Switzerland (Giumaglio Footbridge),

Fig. 11.6 The Rainbow Bridge, Tokyo. *(Photo by Lin.)*

Fig. 11.7 The Tsing Ma Bridge in Hong Kong. *(Photo by Lin.)*

Fig. 11.8 The Pingsheng Bridge Guangdong, China. *(Photo by He.)*

Mozambique (Samora Machel Bridge, 1973), and Nepal (Dhodhara-Chandani Suspension Bridges, 2005). These bridges are generally built in a relatively short span except the Taizhou Yangtze River Bridge in China, which has three main towers and two main spans with a span length of 1080 m, currently are the largest such suspension bridges.

11.3.2 According to Stiffening Girders

Based on the continuity, there are two types of stiffening girders, namely two-hinge or continuous types, as shown in Fig. 11.9. Two hinge stiffening girders are commonly used for highway bridges, while the continuous stiffening girder is often used for combined highway-railway bridges to ensure the continuity between adjacent spans and to secure the smooth operation of the trains (Alampalli and Moreau, 2015). The Akashi Kaikyo Bridge, the longest suspension bridge in the world, was designed with a two hinged stiffening girder system.

11.3.3 According to Suspenders

In suspension bridges, suspenders (or hangers) can be designed as either vertical or diagonal, as shown in Fig. 11.10. Vertical suspenders are more often used in suspension bridges, but diagonal hangers are sometimes used for the sake of increasing the damping and improving the seismic performance of such bridges. For higher stiffness of a cable supported bridge, a combined suspension and cable-stayed cable system can also be used.

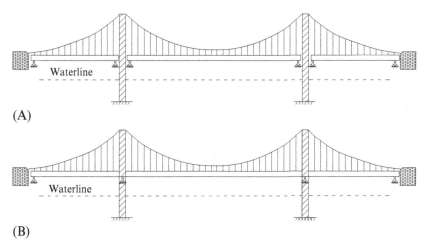

(A)

(B)

Fig. 11.9 Suspension bridge classification according to stiffener girders. (A) Two hinged stiffening girder. (B) Continuous stiffening girder.

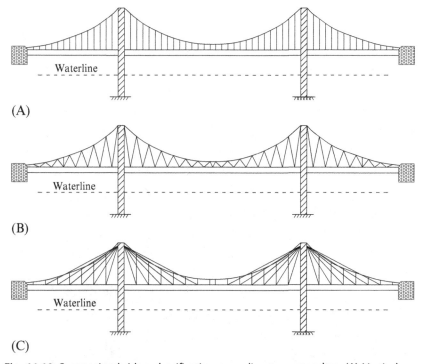

(A)

(B)

(C)

Fig. 11.10 Suspension bridge classification according to suspenders. (A) Vertical suspenders. (B) Inclined suspenders.

11.3.4 According to Anchoring Conditions

Based on anchoring conditions, the suspension bridges can be classified into externally anchored or self-anchored types, as shown Fig. 11.11. For externally anchored suspension bridges, the anchorages need to be built on both ends of the bridges to sustain the tensile forces from the main cable, which is the most common type of suspension bridges. As for self-anchored suspension bridges, the anchorages are not necessary and main cables are connected directly to the stiffening girders. In this case, however, relatively large axial compressive forces need to be carried by the main girder and this should be considered in the design. The San Francisco Oakland Bay Bridge and Konohana Bridge in Osaka (Fig. 11.12) are typical self-anchored suspension bridges.

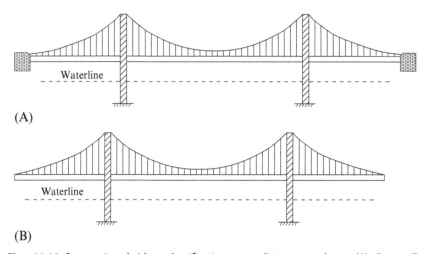

Fig. 11.11 Suspension bridge classification according to anchors. (A) Externally anchored suspension bridges. (B) Self anchored suspension bridges.

Fig. 11.12 The Konohana Bridge (self-anchored suspension bridge) in Osaka, Japan. *(Photo by Lin.)*

11.4 CONFIGURATION

11.4.1 Main Towers

In the longitudinal direction, main towers can be classified into three types; rigid, flexible, and locking types, as shown in Fig. 11.13. Flexible towers are commonly used in long-span suspension bridges, rigid towers for multispan suspension bridges to provide enough stiffness to the bridge, and locking towers occasionally for relatively short span suspension bridges (Okukawa et al., 1999). For towers in suspension bridges, the bucking due to the large compressive forces deserves special attention. In addition, the horizontal seismic performance of the tower is also a concern and seismic analysis should be performed if necessary.

11.4.2 Cables

Chains and eye-bar chains had been used in early suspension bridges, like the Silver Bridge over the Ohio River built in 1928. In 1967, the Silver Bridge collapsed due to the failure of a single eye-bar chain, resulting in the death of 46 people. The Clifton Suspension Bridge (Fig. 11.2) and the South Portland Street Suspension Bridge (Fig. 11.14) in the UK, and Kiyosubashi Bridge (Fig. 11.15) in Japan are also elegant suspension bridges with chain cables.

For modern long-span suspension bridges, the cold drawn and galvanized steel wires are generally used. In general, strands are bundled into a circular shape to form one main cable. Hanger ropes can be steel rods, steel bars, stranded wire ropes, and parallel wire strands, etc. The main cable of the

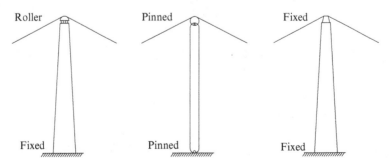

Fig. 11.13 Towers in suspension bridge.

Fig. 11.14 The South Portland Street Suspension Bridge in Glasgow. *(Photo by Lin.)*

Fig. 11.15 The Kiyosubashi Bridge in Tokyo (1928). *(Photo by Lin.)*

Akashi Kaikyo Bridge is made of parallel wire strands covered with polyethylene tubing, as shown in Fig. 11.16.

11.4.3 Stiffening Girders

In suspensions, the often used stiffening girders are I-girders, trusses, and box girders, as shown in Fig. 11.17. I-girders (or plate girders) are simple

Fig. 11.16 Cross-section of the main cable in the Akashi Kaikyo Bridge. *(Photo by Lin.)*

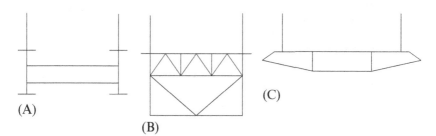

Fig. 11.17 Cross-section of stiffening girder. (A) I-girder. (B) Truss. (C) Box-girder.

in design but has disadvantageous with regard to aerodynamic stability. For modern long-span suspension bridges, trusses, or box girders are more often used.

With the increase of the stiffening girder height, the bridge stiffness will increase, while the deadweight and the bending moment due to temperature

Fig. 11.18 The 25 de Abril Bridge in Lisbon. *(Photo by Lin.)*

variation will also increase. For long-span suspension bridges, the dead-weight takes up a big percentage of the load carrying capacity, and the live load carrying capacity can be increased with the reduction of the dead-weight. Thus, the thinner stiffening girder is often used to reduce the dead-weight. For truss girder, the girder height is generally taken as $1/40$–$1/80$ of the span length. The 25 de Abril Bridge in Portugal and the Great Seto Bridge in Japan are suspension bridges with truss girders, as shown in Figs. 11.18 and 11.19.

11.4.4 Anchorages

Anchorages (or anchors) are important members in suspension bridges because most of the self-weight and other load of the bridge is finally trans-ferred by the cables to the anchorage systems. Inside the anchorages, the cables are spread over a large area to evenly distribute the load and to prevent the damages that may be caused by concentrated cable forces. The safety check should be confirmed to avoid the rotary movement and slippage of the anchorages.

In general, anchorage structure includes the foundation, anchor block, bent block, cable anchor frames, and protective housing. There are two often used anchorages: gravity type or tunnel type anchorage systems, as

Fig. 11.19 The Seto Oohashi Bridge. *(Photo by Taniguchi.)*

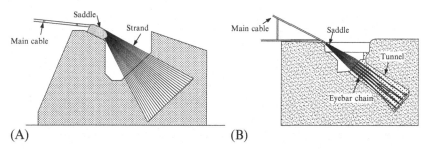

Fig. 11.20 Anchorage systems. (A) Gravity type. (B) Tunnel type.

shown in Fig. 11.20. Gravity anchorage is generally built as huge concrete blocks, which relies on the mass of the anchorage itself to resist the tension of the main cables. This type is commonplace in many suspension bridges. Tunnel anchorage takes the tension of the main cables directly into the ground, and adequate geotechnical conditions are necessary. Inside the anchorages, the cables are spread over a large area to evenly distribute the load and to prevent the cables from breaking free.

11.5 ANALYSIS OF SUSPENSION BRIDGES

11.5.1 Sag and Sag Ratio

Vertical interval (f) of the main cable in the main span is denoted as the sag, and the ratio of sag to span of main cable $(n=f/L)$ is defined as the sag ratio. Sag controls the length and stability of the suspension bridge, and is generally taken between 1/8 and 1/12 (a generally accepted optimum is a 1/10). If the main cables have a flat curve or a low sag ratio, the bridge has more vertical stability but the cable stress is high and strong anchorages are required. On the contrary, a deeper sag reduces cable force, but increases the height of the towers and makes them more susceptible to large forces.

The geometrical shape of a main cable is shown in Fig. 11.21, in which a coordinate system is built at the top of the left tower. The x axis denotes the horizontal location, and y axis represents the cable deflection. Due to the zero moment of the cable,

$$M = M_0 - Hd_y = 0. \tag{11.1}$$

If the deadweight intensity is assumed as ω, the bending moment at the span center can be determined as $M_0 = \omega L^2/8$ considering the simply supported beam theory. For the cable with a sag of f, the horizontal force H can be determined as:

$$H = \frac{\omega L^2}{8f} \tag{11.2}$$

For any point on the main cable, as the bending moment is $M_0 = \omega x(l-x)/2$, thus

$$y = \frac{\omega x}{2H}(L-x) \tag{11.3}$$

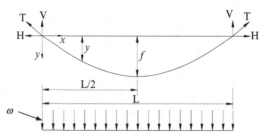

Fig. 11.21 Shape of the main cable.

Taking Eq. (11.2) into consideration, the geometrical shape equation of a main cable can be determined as:

$$y = \frac{4fx}{L^2}(L - x) \tag{11.4}$$

Therefore, if the cables and deadweight of the stiffening girder are uniformly distributed, the main cables generally deform as a parabolic curve.

11.5.2 Analytical Methods for Suspension Bridges

Elastic theory and the deflection theory are generally used for analyses of the global behavior of a suspension bridge system. In both these two methods, the main cable is assumed to be completely flexible with only axial tensile force, the deadweight is uniformly distributed with the main cable deformed like a parabolic. The difference between the two theories is whether the live load is considered for the cable deflection. However, with the development of the computing technology and analysis software; computer based analysis methods, like the finite element methods (FEM), are mainly used for designing suspension bridges.

11.6 SUSPENSION BRIDGE CONSTRUCTION

The construction of the suspension bridge should be organized in an appropriate sequence for the favorable transmission of dead load. In general, the anchorage, the towers, the temporary suspended walkways (called catwalks), the main cables and suspenders, the stiffening girders, are erected in sequence. For steel towers, the tower segments can be erected using crawler cranes for land-based steel towers, and floating cranes for offshore towers. In general, climbing cranes are necessary for erection of the top tower segments. The erection of the stiffened girders is discussed below.

The erection methods for stiffening girder are different for externally anchored and self-anchored suspension bridges. For a self-anchored suspension bridge, some falsework during construction is necessary because the primary cables cannot be anchored until the bridge deck is completed. For externally anchored suspension bridges, the falsework is not necessary for the construction of stiffening girders, and there are two often used erection sequence (ICE, 2008; Okukawa et al., 1999): (1) deck erection starting at mid- main-span and anchorages; or (2) deck erection starting at towers, as shown in Fig. 11.22.

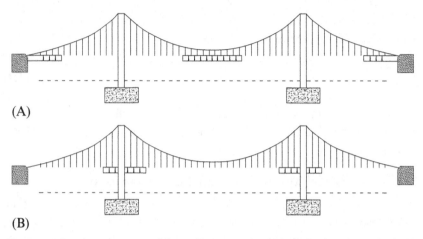

(A)

(B)

Fig. 11.22 Erection sequence of the stiffening girders. (A) Method-1. (B) Method-2.

The main girder is started by lifting a segment at the mid-span by a crane, and then the sections are connected to the hangers. The adjacent girder segments are connected by temporary joints. In the early stages of this construction method, the deflection of the main cable can be large and the bottom flanges cannot be connected. But with the erection of new girder segments, the cable profile becomes close to the design profile, then connection by bolting or welding is carried out. The girder erection at side spans can be carried out simultaneously with the main span to reduce the bending moment at the tower and to adjust the main cable profile. For a suspension bridge with a strong tower, it is possible to erect the girder at side spans after completion of the main span.

For the method-2, the deck erection starts at towers as shown in Fig. 11.22B. The girder erection starts from the main towers until the center of the main span is reached, with the side spans erected either working from the towers or the anchorages, again either concurrently with the main span or subsequently. Similarly, the part-loading of the cable again also causes a pronounced sagging curvature of the deck with open bottom joints at the early stage of erection.

11.7 EXERCISES

1. Describe the structural components of a typical suspension bridge and the internal forces they need to sustain.

2. Classify suspension bridges according to (a) number of spans, (b) continuity of stiffening girders, (c) types of suspenders, and (d) types of cable anchoring.

REFERENCES

Alampalli, S., Moreau, W.J., 2015. Inspection, Evaluation and Maintenance of Suspension Bridges. CRC Press, Boca Raton, FL.

Gerner, M., 2007. Chakzampa Thangtong Gyalpo—Architect, Philosopher and Iron Chain Bridge Builder. Center for Bhutan Studies, Thimphu. ISBN 99936-14-39-4.

Institution of Civil Engineers, 2008. ICE Manual of Bridge Engineering, second ed. Thomas Telford, London.

Okukawa, A., Suzuki, S., Harazaki, I., 1999. Chapter 18: suspension bridges. In: Chen, W.F., Duan, L. (Eds.), Bridge Engineering Handbook. CRC Press, Boca Raton, FL.

Waddell, L.A., 1905. Lhasa and Its Mysteries. John Murray, London. p. 313.

CHAPTER TWELVE

Bridge Bearings and Substructures

12.1 INTRODUCTION

As described earlier, a bridge consists of the superstructure, the substructure, and bearings. Thus, a bearing is a component of a bridge which typically locates between bridge substructures (piers or abutment) and bridge superstructures, playing an important role in the force transmission and in accommodating the deformation caused by temperature variation and the earthquake. A bridge bearing carries the loads or movement in both vertical and horizontal directions from the bridge superstructure and transfers those loads to the bridge piers and abutments. The loads can be live load and dead load in vertical directions, or wind load, earthquake load, etc., in horizontal directions.

The use of a bearing is also to allow controlled movement and thereby reduces the stresses involved. Movement could be thermal expansion or contraction, or deformations caused by creep and shrinkage, or movement from other sources such as seismic activity. The rotations can be caused by traffic or uneven settlement of foundations. In all civil engineering structures, the loads must find a way to the ground. The typical live load transfer path in a girder bridge is shown in Fig. 12.1, in which it can be indicated that the bridge bearing is the transmission node between bridge substructures and superstructures.

However, it should be noted that not all bridges have bearings. For bridges without bearings, the extra forces due to movement restriction must be considered in the design.

Most common types of construction pertaining to substructures are abutment, pier, and foundation. Where appropriate, piers and abutments shall be designed to withstand dead load, erection loads, live loads on the roadway, wind loads on the superstructure, forces due to stream currents, floating ice and drift, temperature and shrinkage effects, lateral earth and water pressures, scour and collision, and earthquake loadings (Caltrans. 2000).

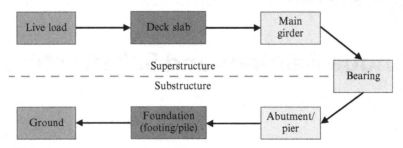

Fig. 12.1 Live load transfer path in bridge structures.

12.2 BEARINGS

12.2.1 Bridge Bearing Classification According to Degrees of Freedom

Basically, the bearings can be classified depending upon degree of freedom, or manufacturing material.

There are six possible independent degrees of freedom at a support, consisting of three components of translation and three angles of rotation. A bearing may permit movement in any of these six degrees of freedom or in none. During the structural design of the bridge girders, each support point is idealized in a specific manner by the design engineer. The bearing has to fulfill this assumption (Indian Railway Institute of Civil Engineers, 2006).

But basically, the bearings can be classified into the following three types: (1) fixed bearing or hinged bearing, which allows rotation, but not other forms of movement; (2) moveable bearing such as a roller bearing, which allows the movement in one direction and rotation, but not movement in other directions; or (3) omni-directional moveable bearing, such as the rocker bearing. Another type of mechanical bearing is the fixed bearing, which allows rotation, but not other forms of movement.

12.2.2 Bridge Bearing Classification According to Materials

There are two types of bridge bearing according to the materials: metal bearing and rubber bearing.

12.2.2.1 Metal Bearing

Metal bearings were used widely in the early stage, and different types of metal bearings were developed according to the design requirements by considering durability and maintenance.

(1) Line bearing

The image of a line bearing is shown in Fig. 12.2. The contacting line between the upper plate and the lower round surface offers rotational capacity as well as sliding. This type of bearing is mainly used in short-span bridges.

(2) Plane bearing

Plane bearings are the simple type of bearings usually consist of a low friction polymer, polytetrafluoroethylene, sliding against a metal plate. They do not accommodate rotational movement in the longitudinal or transverse directions and only resists loads in the vertical direction.

(3) Roller bearing

Roller bearings are movable bearings, which allow for horizontal movement and maintain low friction by using single or multiple rollers, for hinged bearing or spherical bearings. There are several types of roller bearings, such as single roller, multiple roller, and roller bearing with gears, as shown in Fig. 12.3. A picture of a roller bearing is shown in Fig. 12.4.

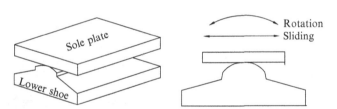

Fig. 12.2 Line bearing.

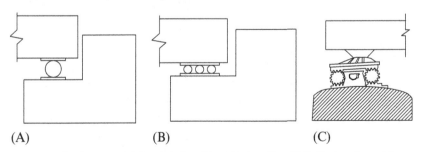

Fig. 12.3 Roller bearing. (A) Single roller. (B) Multiple roller. (C) Roller bearing with gear arrangement.

Fig. 12.4 Roller bearing and pin bearing. *(Photo by Lin.)*

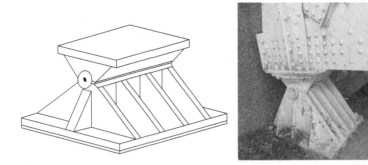

Fig. 12.5 Pin bearing. *(Photos by Lin.)*

(4) Pin bearing

A pin bearing is also called as a hinged bearing, where a steel pin is inserted between the upper and lower shoes allowing rotation but no translational movements. The image of a pin bearing is shown in Fig. 12.5. In general, caps are needed at both ends of the pin to keep the pin from sliding off the seats and to resist uplift loads if required. The upper plate is connected to the sole plate by using bolting or welding.

(5) Pivot bearing

Pivot bearings, as shown in Fig. 12.6, are fixed bearings with a concave upper shoe and a convex lower shoe. They can rotate in all directions and are usually used in truss bridge, suspension bridge, and curved bridges. When combined with sliding elements, movement can be provided.

(6) Rocker bearing

A rocker bearing, as shown in Fig. 12.7, has a great variety as a type of expansion bearing. It typically consists of a pin at the top that accommodates rotations, and a curved surface at the bottom that accommodates the translational movements. In general, rocker bearings are primarily used in steel bridges.

(7) Pendel bearing

Pendel bearings are mainly used for cable-stayed bridges, to resist the negative (or uplift) reaction forces. An eye bar is used to the superstructure and substructure by pinned connection at both ends, as shown in

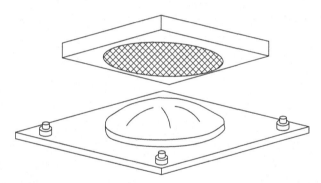

Fig. 12.6 Pivot bearing.

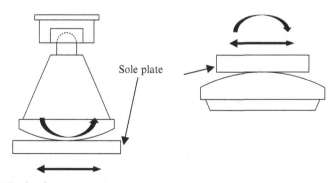

Sole plate

Fig. 12.7 Rocker bearing.

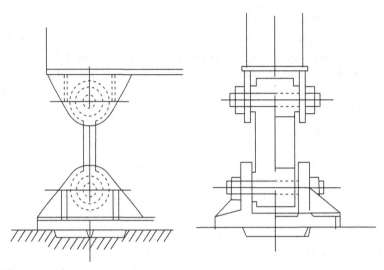

Fig. 12.8 Pendel bearing (Tachibana, 2000).

Fig. 12.8. The longitudinal movement is permitted by the inclination of the eye bars, and there is no resistance in transverse direction. Thus, this type of bearings should be used together with other type of bearings which are capable of resisting the horizontal reaction forces (Toma et al., 2005).

12.2.2.2 Rubber Bearing

(1) Elastomeric

Elastomeric bearings (Fig. 12.9) enable the deck to translate and rotate, and resist a certain amount of loads in the longitudinal, transverse, and vertical directions. In general, they are very flexible in shear but very stiff against volumetric change. Steel plates (or sometimes fiberglass) are typically used to reinforce the pad in alternate layers to

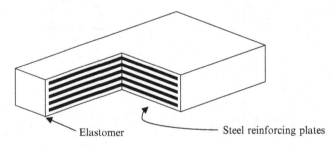

Elastomer   Steel reinforcing plates

Fig. 12.9 Elastomeric bearing.

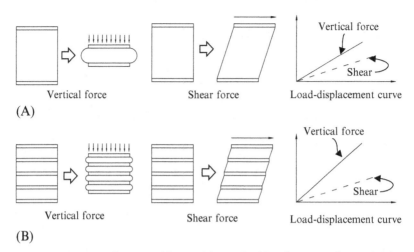

Fig. 12.10 Properties of a unit rubber and layered rubber (Toma et al., 2005). (A) Unit rubber. (B) Layered rubber.

prevent it from "bulging" under high load, allowing it to resist higher vertical compression. The difference between a unit rubber and a layered rubber is shown in Fig. 12.10, indicating that the use of steel plates in layered rubbers can suppress the vertical of the rubber but affect little on the shear deformation (Toma et al., 2005). During an earthquake, the flexibility absorbs horizontal seismic energy and is ideally suited to resist earthquake actions. In Japan, the elastomeric rubber bearings have used more widely since the 1995 Kobe Earthquake.

The following equation can be used to determine the stiffness of rubber bearing in vertical direction:

$$K_c = \frac{(3 + 6.5S^2)GS_s}{\sum t_e} \tag{12.1}$$

$$S = \frac{A_s}{\{2(a + b)\}} \tag{12.2}$$

where G is the shear modulus of rubber, A_s is the cross-sectional area of the steel plate $(a \times b)$, t_e is the thickness of each rubber layer, a is the length of steel plate, and b is the width of the steel plate.

To determine the stiffness of rubber bearing in horizontal direction (Fig. 12.11), the following equation can be used:

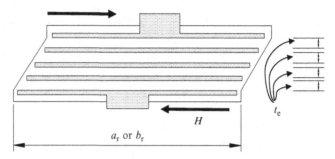

Fig. 12.11 Rubber bearing under horizontal force.

$$K_r = \frac{GA_r}{\sum t_e} \qquad (12.3)$$

where A_r is the cross-sectional area of the rubber bearing $(a_r \times b_r)$.

When subjected to the maximum reaction force R_{max} and maximum horizontal deformation δ, the maximum compressive stress is determined as (Fig. 12.12):

$$\sigma_{max} = \frac{R_{max}}{\{b(a - \delta)\}} \qquad (12.4)$$

(2) Pot bearing

Pot bearings have a cylindrical elastomeric pad confined with a pot, and they are suitable for high loads, displacements, and rotations. Special components, such as bolt connections (shear studs) are generally used for connecting the bearing itself with bridge superstructure or substructure.

Fig. 12.12 Elastomeric bearing. *(Photos by An.)*

(3) Seismic isolation bearing

High damping rubber bearings between the bridge super- and substructures are used to increase the period and damping of the original structure system, by reducing the shear stiffness of the bearing. As a result, to achieve the purpose of reducing the energy delivered to the superstructure and the requirement of anticipated shock resistance. The bridge rubber shock isolation bearing is not only to deliver the load and bear the deformation but also to establish proper seismic isolation and improve the seismic performance of the bridge structures during the service stage.

In Japan, large revision and improvement was conducted on the bridge design specification after the 1995 Kobe Earthquake. Before this earthquake, 80% of the bridges were using metal bearings, while only 20% used the rubber bearing. However, after the earthquake, the 80% of the new bridges began to use the rubber bearing, and only 20% of the new bridges use the metal bearing. This is because the metal bearings do not have extra deformation capacity beyond the ultimate strength. However, the effectiveness of such design method still needs further investigations (Tachibana, 2000).

12.2.3 Bearing Design Considerations

Selection of bearing should be made according to the size of the bridge, magnitude of predicted (upward or downward) vertical and horizontal loads, translational and rotational movements, etc. In addition, movements of the bridge structure enabled by bearings are in relation to a preset allowance integrated in the bridge structure to accommodate elongation provided by expansion joint. Other factors, such as the life cycle cost (initial cost, inspection, and maintenance cost), environment (corrosion/temperature range), easy replacement or reset, etc., should also be taken into consideration.

12.3 ABUTMENTS

An abutment is a very important part of a bridge as it transfers the loads from the superstructure to the earth. In general, abutments are concrete structures composed of bridge seat, parapet, main body, and footing. Abutments need to have sufficient capacity to resist overturning, sliding, settlement. Abutments are designed to withstand earth pressure, the weight of the abutment and bridge superstructure, live load on the superstructure or approach fill, wind forces and longitudinal forces when the bearings are

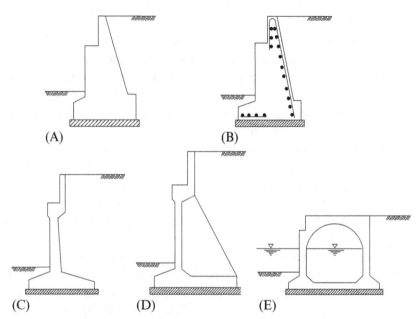

Fig. 12.13 Types of abutments. (A) Gravity type. (B) Semigravity type. (C) Inverted T-type. (D) Counterforted type. (E) Portal type.

fixed, and longitudinal forces due to friction or shear resistance of bearings. The design shall be investigated for any combination of these forces which may produce the most severe condition of loading. There are several types of bridge abutments, such as the gravity type, semigravity type, inverted T-type, counterforted type, and portal type, as shown in Fig. 12.13.

The procedure of selecting the most appropriate type of abutments can be based on the following consideration (AASHTO, 1996):

1. Construction and maintenance cost
2. Cut or fill earthwork situation
3. Traffic maintenance during construction
4. Construction period
5. Safety of construction workers
6. Availability and cost of backfill material
7. Superstructure depth
8. Size of abutment
9. Horizontal and vertical alignment changes
10. Area of excavation
11. Esthetics and similarity to adjacent structures
12. Previous experience with the type of abutment

13. Ease of access for inspection and maintenance
14. Anticipated life, loading condition, and acceptability of deformations

12.4 PIERS
12.4.1 General

Piers are substructures located at the ends of bridge spans at intermediate points between the abutments. The function of the piers is as follows: to transfer the superstructure vertical loads to the foundation and to resist all horizontal and transverse forces acting on the bridge. Piers are generally constructed of masonry or reinforced concrete. Since piers are one of the most visible components of a bridge, the piers contribute to the esthetic appearance of the structure. They are found in different shapes, depending on the type, size, and dimensions of the superstructure and also on the environment in which the piers are located.

12.4.2 Pier Types

Piers may be grouped into bent types, inversed T-types, portal types, column types, and high pier types. Bents are a bridge support system consisting of one or more columns supporting a single cap. Columns are defined as a single support member having a ratio of clear height to maximum width of 2.5 or greater. The columns may be supported on a spread- or pile-supported footing, or a solid wall shaft, or they may be extensions of the piles or shaft above the ground line. Typical shapes of piers commonly used in practice are as shown in Fig. 12.14. They can be inversed T-type, portal type. Inversed T-type piers for river bridges are provided with semicircular cut-waters to facilitate and streamlined flow and to reduce the scour. Solid piers can be of mass concrete or of masonry for heights of up to 6 m and spans

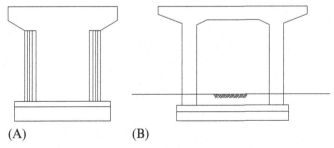

(A)   (B)

Fig. 12.14 Types of piers. (A) Inverted T-type. (B) Portal type.

up to about 20 m. Portal type piers are increasingly used in urban elevated highway applications, as it provides slender substructure with open and free-flowing perception to the motorists using the road below. It is also used for river crossings with skew alignment.

12.5 FOUNDATIONS

12.5.1 General

A foundation is the part constructed under the abutment and pier and over the underlying soil or rock. Foundation is one of the most important structural part of bridge superstructures, which receives the load from the piers and abutments and transfer it to the soil. Foundations shall be designed to support all live and dead loads, and earth and water pressure loadings. Different types of foundations in common use are shown in Fig. 12.15.

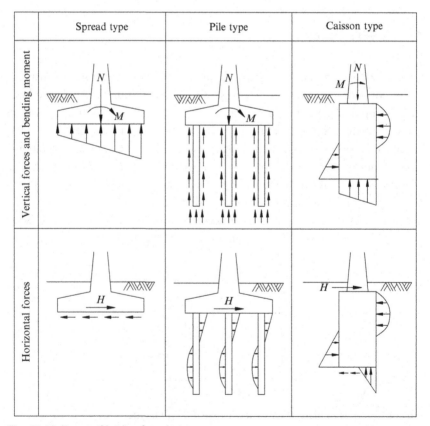

Fig. 12.15 Types of bridge foundations.

The controlling factors in selecting the shape of the foundation are: the base dimensions of pier or abutment, the ease with which the foundation can be sunk, cost, considerations of tilt and shift, ease of sinking, and the magnitude of the forces to be resisted by the foundation.

12.5.2 Foundation Types and Capacity

The selection of the foundation system for a particular site depends on many considerations, including the nature of subsoil, location where a bridge is proposed to be constructed, i.e., over a river, road, or a valley, and the scour depth. The different types of foundations for bridges are divided into two types: shallow foundation and deep foundation. Common forms of construction pertaining to shallow foundations are spread foundations and raft foundations. Common forms of construction pertaining to deep foundations are pile foundations, caisson foundations, and well foundations.

(1) Selection of Foundation Type

Selection of foundation types shall be based on an assessment of the magnitude and direction of loading, depth to suitable bearing materials, evidence of previous flooding, potential for liquefaction, undermining or scour, swelling potential, frost depth, and ease and cost of construction.

(a) *Spread foundations*: They may be provided for small bridges when bed is of rock or soil having good bearing capacity.

(b) *Pile foundations*: They consist of relatively long and slender members, called piles which are used to transfer loads through weak soil or water to deeper soil or rock strata having a high bearing capacity. They are also used in normal ground conditions for elevated road ways. The analysis and the design of all the components of a bridge particularly with reference to the bridge substructure can become a very lengthy and laborious task if the calculations are attempted manually.

(c) *Caisson foundations*: They are the most common types of foundations for both road and railway bridges. Such foundations can be sunk to great depths and can carry very heavy vertical and lateral loads. Caisson foundations are suitable when depth of water in the river is more and quality of bed soil is good. Caissons may be of circular, rectangular, or any other shape.

(2) Foundation Capacity

Foundations shall be designed to provide adequate structural capacity, adequate foundation bearing capacity with acceptable settlements, and acceptable overall stability of slopes adjacent to the foundations. The tolerable level of structural deformation is controlled by the type and span of the superstructure.

(3) Soil, Rock, and Other Problem Conditions

Geologic and environmental conditions can influence the performance of foundations and may require special consideration during design. To the extent possible, the presence and influence of such conditions shall be evaluated as part of the subsurface exploration program.

12.6 EXERCISES

1. Describe the main functions of bridge bearings.
2. Describe the classifications of bridge bearings.
3. List up the factors that should influence the selection of abutments in bridge design.
4. Describe possible classifications of bridge foundations.

REFERENCES

AASHTO, 1996. AASHTO Standard Specification.
Caltrans, 2000. Bridge Design Specifications. Substructures. Caltrans, Sacramento, CA (Section 7).
Indian Railway Institute of Civil Engineers, 2006. Bridge Bearings.
Tachibana, Y., 2000. In: Nakai, H., Kitada, T. (Eds.), Bridge Engineering. Kyoritsu Shuppan Co., Ltd., Tokyo, Japan.
Toma, S., Duan, L., Chen, W.F., 2005. Chapter 25: bridge structures. In: Chen, W.F., Lui, E.M. (Eds.), Handbook of Structural Engineering. second ed. CRC Press, Boca Raton, FL.

CHAPTER THIRTEEN

Inspection, Monitoring, and Assessment

13.1 INTRODUCTION

With a large share of infrastructure systems, aged bridges are facing an increasing risk of failures, due to the deterioration of structural members caused by corrosion, fatigue, cracks, concrete alkali-silica reaction, and the rise of traffic load (occasional overloading), etc. In addition, apart from these inevitable service reasons, other extreme events resulting from accidents or natural disasters such as ship-collision, flood, hurricane, and earthquake, etc., also threaten bridges' safety. The occurrence probabilities of extreme conditions directly correlate with service time. Thus, in order to increase reliability of aged bridges, engineers are considering different ways to settle these issues associated with aged bridges. In order to maintain properly the aged and deteriorated bridges, appropriate inspection, monitoring, and assessment are needed.

13.2 BRIDGE INSPECTION

13.2.1 Objectives of Inspection

Bridge inspection is an essential element of any bridge management system particularly for aged and deteriorated bridges and a path way to condition rating. The validity of condition assessment relies heavily on the quality of the inspection. A variety of bridge inspections may be required on a bridge during its service life (Fig. 13.1).

The objectives of bridge inspection are:
1. Secure and guarantee bridge safety.
2. Record bridge condition systematically and periodically.
3. Find abnormalities and predict damage.
4. Provide data necessary to make decisions with regard to repairs, strengthening, and replacement.

Bridge Engineering
http://dx.doi.org/10.1016/B978-0-12-804432-2.00013-X

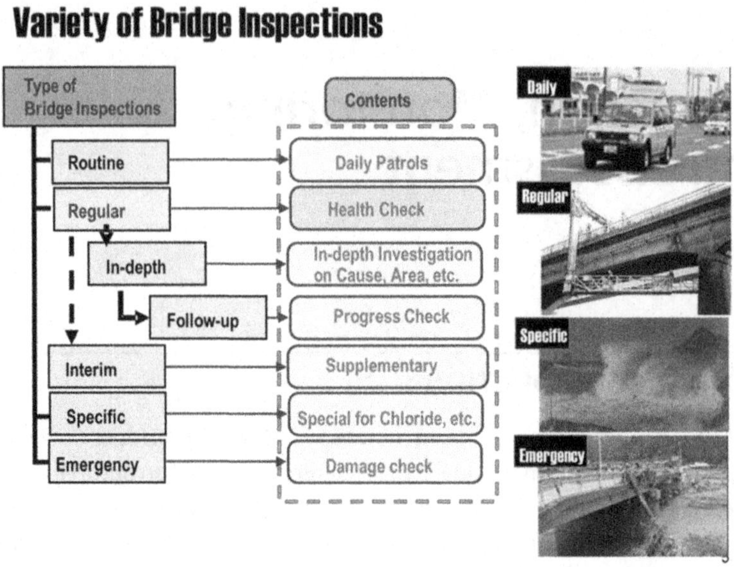

Variety of Bridge Inspections

Fig. 13.1 Variety of bridge inspection. *(Courtesy of MLIT.)*

5. Set up a reasonable maintenance plan based on the data of bridge inspection results considering life cycle cost.
6. Analyze accumulated bridge inspection results for the improvement of design and construction of bridges.

Bridge inspection is the use of techniques to determine the condition of a bridge, with the user of these techniques being the Bridge Inspector. Good bridge inspection consists of five steps (FHWA, 2012):

1. Planning of the inspection to ensure that attention is given to each bridge component in accordance with its importance,
2. Preparation according to the inspection plan,
3. Careful and systematic observations,
4. Complete and accurate recording of significant observations,
5. Assessment of observations to determine condition ratings.

13.2.2 Fundamentals of Inspection

Firstly, the inspector should conduct a cursory visual inspection of the entire bridge looking for indications of problems. Next, the inspector should conduct a hands-on visual inspection of the bridge parts taking into account any indications of problem found during the cursory inspection. During the hands-on visual inspection, the inspector should look for signs

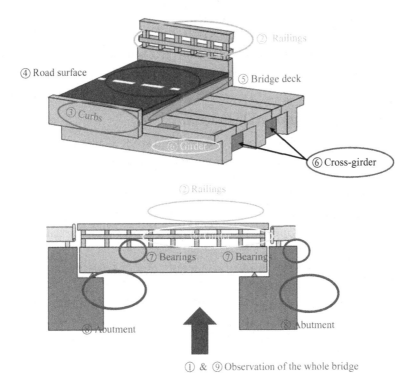

Fig. 13.2 Inspection order.

of deterioration that will need a physical examination. In general, the inspector should start at the top of the bridge and work one's way down the load paths. This will help the inspector from missing any parts of the structure.

One inspection order could be (Fig. 13.2):

① Observation of the whole bridge
② Observation of railings
③ Observation of curbs
④ Observation of road surface
⑤ Observation of bridge deck
⑥ Observation of girder/cross-girder
⑦ Observation of bearings
⑧ Observation of abutments
⑨ Observation of the whole bridge (final check)

As examples of typical damage, damage to steel members could be: ① Corrosion, ② Fatigue crack, ③ Loosening or falling of bolts, ④ Fracture, ⑤

Deterioration of anticorrosion function; in addition to this, damage to concrete members could be: ⑥ Crack, ⑦ Spalling/reinforcing steel exposure, ⑧ Leakage/efflorescence, ⑨ Spalling off, ⑩ Damage to repair members, ⑪ Crack in floor system, and ⑫ Separation.

13.2.3 Basic Methods of Inspections

Basically, there are two methods used to inspect a bridge: visual and advanced inspection techniques. In general, visual inspections reveal areas that require further investigations.

Types of visual inspections are:

1. Cursory inspection
   - Involves reviewing the previous inspection report and visually examining the members.
   - Involves a visual assessment to identify obvious defects.
2. "Hands-on" inspection (Fig. 13.3)
   - Requires the inspector to visually assess all defective surfaces at a distance no further than an arm's length.
   - Surfaces are given close visual attention to quantify and qualify any defects.

Fig. 13.3 Hands-on inspection. *(Photo by Yoda.)*

It is important to document the inspection findings. Documentation should include:
- An inspection report and notes
- And lots of photographs

Photographs should be a minimum of 10 pictures with additional pictures of problem areas.

Minimum required photographs:
1. Near approach looking at the bridge
2. Far approach looking at the bridge
3. Bridge deck and railing
4. Underside of the deck and beams
5. Upstream looking downstream at the bridge
6. Downstream looking upstream at the bridge
7. Looking upstream from the bridge
8. Looking downstream from the bridge
9. Near side substructure
10. Far side substructure

13.2.4 Types of Bridge Inspections

1. Initial (inventory) inspection

Initial inspections are performed on new bridges or when existing bridges have no appropriate database. This inspection provides a basis for all future inspections or modifications to a bridge.

Inventory inspections provide structural inventory and appraisal data along with bridge element information and baseline structural conditions (Rashidi and Gibson, 2012).

Inventory inspections usually start in the office with the construction plans and route information then proceed to the field for verification of the as-built conditions. Initial defects which might not have been present at the time of construction are noted. Changes in the condition of the site, such as erosion, scour, and regrading of slopes should also be noted.

2. Routine inspection

The routine inspection is a diagnostic method with the greatest potential and is generally based on direct visual observation of a bridge's most exposed areas. It relies on subjective evaluations made by the bridge inspectors. During a routine inspection, no significant structural defect is expected and the work recommended falls within the scope of maintenance. A routine inspection must be planned in advance to facilitate the best assured conditions (e.g., weather conditions and traffic) that may permit detection of defects.

3. Detailed inspection

    Easy and fast nondestructive in situ tests are performed in detailed inspection in addition to direct visual observation as a way of exploring every detail that may potentially lead to future problems. There is a possibility that special means of access may be used if such is considered indispensable (Branco and Brito, 2004).

4. Special inspection

    This could be undertaken to cover special conditions such as occurrences of earthquakes, unusual floods, passage of high intensity loading, and heavy deterioration, etc. These inspections should be supplemented by testing as well as structural analysis, as shown in Fig. 13.4. For this reason the inspection team should include an experienced bridge design engineer (Rashidi and Gibson, 2012) (Table 13.1).

Fig. 13.4 Vibration test of a Truss Bridge. *(Photos by Yoda.)*

Table 13.1 Features of Maintenance Manual

Items	Features of Maintenance Manual
Maintenance plan	• General items about maintenance • Project overview, organization operation plan • Law and regulations for maintenance
Classification of structural system	• Standards on structural classification for effective maintenance, classification of each bridge • Composition of database system for bridge maintenance • Classification of bridge, approach bridge • Classification of members of the whole bridge
Inspection system	• Inspection plan, inspection details, inspection methods for each member • Inspection plan, access, detailed inspection, inspection methods, and action

Table 13.1 Features of Maintenance Manual—cont'd

Items	Features of Maintenance Manual
Assessment of condition and safety	• A comprehensive method for relative assessment, durability test, load-carrying capacity assessment, and safety test after inspection, • Assessment of the final grade of the whole bridge • Assessment methods and procedures
Emergency maintenance	• Emergency measures for unexpected conditions that occurred during bridge service • Emergency inspection (earthquake, strong wind, collision of vehicles and ships, fire) • Safety management of workplace
Preventive maintenance	• Prevent damage of major members with preventive maintenance methods • Life cycle cost of bridge • Preventive maintenance of each member
Repairs and strengthening	• Repairs/strengthening for damage and reduced durability
Maintenance facilities	• Maintenance measures of inspection facilities for maintenance

13.3 BRIDGE MONITORING

13.3.1 Objectives of Monitoring

Monitoring system has the following characteristics:

1. Identify design weaknesses
2. Ensure optimal risk follow-up with existing instrumentation
3. Complete follow-up actions with consistent inspections and investigations
4. Identify early signs of vulnerabilities and quantify their progression
5. Optimize preventive maintenance based on inspection results
6. Optimize inspection and maintenance management with a detailed and customized inspection and maintenance system
7. Maintenance that requires high cost.

From the monitoring characteristics, monitoring objectives lead to:

1. Number of measurement items necessary to understand the structural behavior and follow its evolution,
2. Required durability and precision for each measurement item,
3. Required reliability: 24 h a day all year round is the standard,

4. Data acquisition capabilities, data-processing capabilities, data analysis capabilities, and data storage and archiving,
5. Data transmission protocols: remote automated measurements feeding data directly into the system, versus field measurements performed by staffs on a periodic basis,
6. Cost of monitoring system (including supply, installation, and operation),
7. Comparison to benefits obtained with the system,
8. Comparison of alternative techniques.

13.3.2 Fundamentals of Monitoring

The overall framework of monitoring should include: (1) networked sensor arrays, (2) a high-performance database, (3) computer vision applications, (4) tools of data analysis and interpretation in light of physics-based models, (5) visualization allowing flexible and efficient comparison between experimental and numerical simulation data (Fig. 13.5), (6) probabilistic modeling, structural reliability, and risk analysis, and (7) computational decision theory (Elgamal et al., 2003).

13.3.3 Database Research

The complexity of data sources (including real-time sensor and video streams, and the output of physics-based and statistical models), and the need

Fig. 13.5 Monitoring of corrosion environment due to sea breeze. *(Photo by Yoda.)*

to perform advanced real-time and off-line analyses requires a high-performance computational infrastructure.

13.3.4 Sensor Network

The sensor network consists of a dense network of heterogeneous sensors (e.g., strain gages, accelerometers, cameras, potentiometers, etc.). In addition, the network must be easy to deploy, scalable—allowing for progressive deployment over time, and must allow for local processing and filtering of data, remote data collection, and accessibility and control (Elgamal et al., 2003).

13.3.5 Computer Vision

Visualization is often the first step in data exploration, enabling scientists and decision makers to exploit the pattern recognition capabilities of the human visual system. Visualizations of sensor measurements, features obtained from measurements, and simulation results provide visual interpretations of infrastructure status and behavior.

13.3.6 Data Analysis

Data analysis includes tasks aimed at evaluating, calibrating, and applying several appropriate approaches for detecting small structural changes or anomalies and quantifying their effects up to the decision making process (Fig. 13.6).

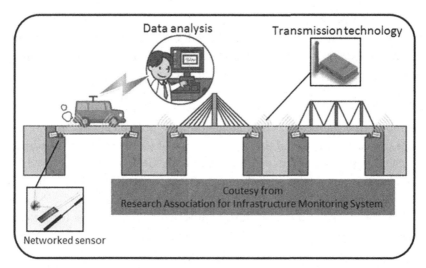

Fig. 13.6 Bridge monitoring system. *(Courtesy from Research Association for Infrastructure Monitoring System.)*

13.4 STRUCTURAL ASSESSMENT

13.4.1 Objectives of Assessment

A structural assessment is normally the consequence of the detection of a major structural or functional deficiency during a routine or detailed inspection. The expected results from this inspection will be: the characterization of the structural shortcomings, the remaining service life estimation by using degradation mathematical models, and also evaluation of its present load-carrying capacity. The static and dynamic load tests and laboratory tests can be valuable complements to the information collected in situ (Rashidi and Gibson, 2012),

- Recognize common causes of defects in concrete/steel bridges.
- Plan an inspection/testing program, and interpret the results.
- Carry out an assessment (with supervision) to performance requirements.
- Appreciate various repair/maintenance options.
- Improve their appreciation of the avoidance of deterioration problems in the design of new bridges and the repair of existing bridges.

13.4.2 Structural Assessment

Most countries adopt component-oriented design and evaluation techniques to verify the safety of structures. Systematic effects of the structures are ignored in the conventional design codes. Namely, safety of every member in the structures is required to ensure overall safety of structures. Some requirements of current design codes, like serviceability limit state requirements, demand overall performance of bridges. However, it does not guarantee collapse-proof safety of the structures. As for an aged bridge, damage of members to some extent is inevitable but this is only acceptable as long as it does not cause collapse of the bridge or lose of its serviceability. Although this component-oriented design method has been successfully used for decades, it is not able to provide a quantitative conclusion on overall safety of aged bridges.

This criterion covers design requirements (Fig. 13.7) for bridges to ensure they are functional, robust, durable, and safe. It defines procedures and provides the technical criteria required to design new structures and to widen and strengthen existing bridges for service life and durability.

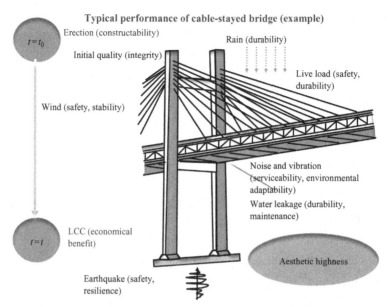

Fig. 13.7 Typical example of performance.

The following risk assessment covers the design requirements and load-carrying capacity assessment requirements for both new and existing bridges:
1. Environmental hazards: earthquake, strong winds, soil erosion, and flooding,
2. Accidents and incidents: ship or vehicle collision, overloading, fire …
3. Degradation of materials: fatigue, corrosion, cracking, aging …
4. Structural deficiencies: design and construction deficiencies, lack of maintenance.

13.5 EXERCISES

1. Describe the classifications and basic methods of bridge inspections.
2. Why bridge monitoring is important, and describe the overall framework of monitoring.
3. List up the factors that should influence the selection of abutments in bridge design.
4. Describe possible classifications of bridge foundations.

REFERENCES

Branco, F.A., Brito, J., 2004. Handbook of Concrete Bridge Management. American Society of Civil Engineers, Reston, VA.

Elgamal, A., et al., 2003. Health monitoring for civil infrastructure. In: 9th Arab Structural Engineering Conference (9ASEC), At Abu Dhabi, United Arab Emirates.

FHWA, 2012. Bridge Inspector's Reference Manual (BIRM). Federal Highway Administration (FHWA), Washington, DC.

Rashidi, M., Gibson, P., 2012. A methodology for bridge condition evaluation. J. Civ. Eng. Archit. 6 (9 (Serial No. 58)), 1149–1157.

APPENDIX

Case Study: Railway Bridge Monitoring Using Fiber-Optic Sensors

Liam J. Butler

University of Cambridge, Cambridge, United Kingdom

Project Background

The Stafford Area Improvements Program was a £250 million rail infrastructure upgrade and redevelopment project in the United Kingdom. It aimed to reduce congestion, improve maintainability, and increase train speeds on the West Coast Main Line (WCML) in the UK by creating a grade-separated junction. As part of the project, 11 new bridges were constructed. As part of a collaborative research project undertaken by the Center for Smart Infrastructure and Construction (CSIC) at the University of Cambridge and the project partners Atkins, Laing O'Rourke, Network Rail, and VolkerRail, an advanced integrated fiber-optic monitoring system was tested for deployment on a bridge structure.

The bridge is a 26.8 m Network Rail type "E" half-through steel rail bridge with reinforced concrete composite deck carrying two train lines over several existing lines of the WCML. The bridge has skewed ends with a skew angle of 23 degrees. The overall project began in 2013 and was completed in the fall of 2016. In addition to monitoring the bridge superstructure, fiber-optic sensors were installed in several prestressed concrete rail sleepers that were installed on the bridge in order to study the load transfer mechanisms from the train axles, through the railway ballast and onto the bridge deck. Installation of the monitoring system began in May 2015, and readings were taken at critical stages of the construction program until the bridge was completed in Apr., 2016. The completed bridge is depicted in Fig. 13.8.

Fig. 13.8 Completed intersection bridge 5. *(Photo courtesy of L.J. Butler, used with permission.)*

Monitoring Objectives

The primary focus of this study was to use sensor data as a means of tracking the entire load history right from the beginning of the bridges construction. One of the main objectives involved developing robust sensor packaging systems and installation methods for deployment as part of bridge construction processes. Another main objective of this project was to carry out model validation. Sensor data were used to help understand the structural behavior of these types of bridges under permanent and live loading. Finite element model updating that included a detailed analysis of time-dependent concrete properties such as creep and shrinkage was also undertaken as part of this study. The final objective was to develop the framework for a "structural health passport" that would be based on the sensor data acquired prior to completion of the structure. This passport could then be used as a quantitative baseline of structural performance on which to base future monitoring results and long-term asset management decisions.

Sensor Technology

Fiber-optic sensors were used as part of this study. The sensors measured strain and were based on fiber Bragg gratings (FBGs). As compared with traditional electrical-based strain sensors, FBGs have several advantages in that they record strain to higher levels of precision, are noncorrosive and nonconductive, are electromagnetically inert, and multiple FBGs can be created along a single fiber. Bragg gratings are periodic alterations in the refractive index of the optical fiber core, and they behave as selective mirrors allowing specific wavelengths of light (Bragg wavelength) to pass through and others to be reflected. As the fiber is strained, the shift in the Bragg wavelength is measured and converted to a strain change. However, FBGs are also sensitive to temperature changes, and therefore, separate temperature measurements must also be recorded in order to isolate the effects of mechanical or true strain in the sensor. Eq. (13.1) may be used to convert wavelength shifts in a specifically tuned Bragg grating to mechanical strain (i.e., compensated for temperature effects). This equation assumes that both strain and temperature are recorded. However, in applications where temperature changes are relatively small (i.e., during the passage of a fast-moving train), temperature effects may be ignored.

$$\Delta\varepsilon_M = \frac{1}{k_\varepsilon}\left[\left(\frac{\Delta\lambda}{\lambda_0}\right)_S - k_T\frac{\left(\frac{\Delta\lambda}{\lambda_0}\right)_T}{k_{T_T}}\right] - \alpha_{\text{conc}}\frac{\left(\frac{\Delta\lambda}{\lambda_0}\right)_T}{k_{T_T}} \qquad (13.1)$$

where $\Delta\varepsilon_M$ is the change in mechanical strain, $(\Delta\lambda/\lambda_0)_S$ is the change in relative wavelength of the strain sensor, $(\Delta\lambda/\lambda_0)_T$ is the change in relative wavelength of the temperature sensor, k_ε is the gage factor (typically 0.78), k_T is the constant for the FBG temperature compensating sensor, k_{T_T} is the change of the refractive index of glass, and α_{sub} is the linear coefficient of thermal expansion of the substrate (°C).

The FBG strain and temperature sensors were installed on the steel girders and crossbeams using a structural adhesive. FBG cables were manufactured with an additional 1 mm thick GFRP coating to provide additional robustness during installation and to improve long-term durability. Fig. 13.9 shows the installation of the FBG sensor cables along the bottom flange of the crossbeams.

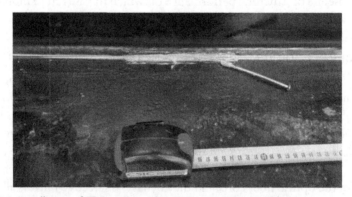

Fig. 13.9 Installation of FBG strain and temperature sensor cables.

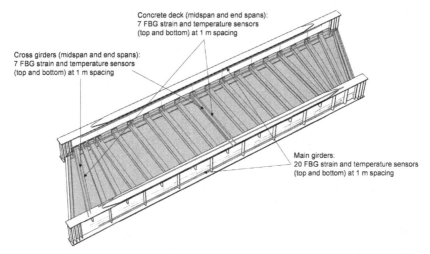

Fig. 13.10 Fiber-optic sensor layout on a bridge superstructure.

FBG sensors were installed in both main girders along the top and bottom flanges and along the top and bottom flanges of the midspan and end (skewed) span crossbeams. The reinforced concrete deck consisted of a two-layer grid of reinforcing steel. FBG sensors were installed transversely along the top layer of reinforcing steel at the locations of the instrumented midspan and end (skewed) span crossbeams. Fig. 13.10 depicts the sensor array locations installed on the bridge superstructure.

Sample Monitoring Results

Measurements were taken throughout the construction stages including during the casting of the concrete deck, over a period of 10 days during the concrete curing, after installation of a temporary haul road, and after the bridge was completed with the ballast, sleepers, rails, and overhead line equipment installed. Readings were also recorded of the bridge response under live train loading (4 months following the bridge being opened to commercial trains). These results are presented in Fig. 13.11.

Note that the readings have assumed the baseline strain to be the strain recorded just prior to the train passing over the bridge. Based on these readings, the load share ratio between each girder can be determined. Based on the top FBG readings, the west main girder carries approximately three times the load as the east main girder. Based on the bottom girder readings, this ratio becomes approximately 3.5:1. The data may also be used to determine the number of train axles and the speed of the passing train based on the time between strain peaks corresponding to the axles and axle spacing. The train

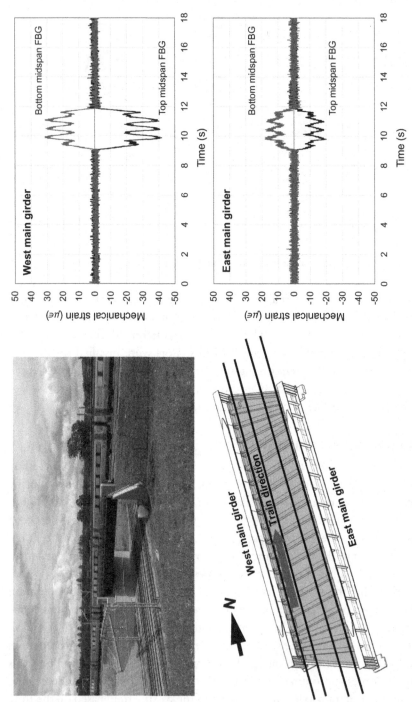

Fig. 13.11 Strain response of main girders during the passage of a commercial train.

recorded in Fig. 13.11 had five separate bogies and was estimated to be traveling at 105 km/h while it passed over the bridge. In this way, a bridge monitoring system can serve additional functions besides monitoring the strain evolution of a structure with time.

Additional information and monitoring results from this project may be found in Further Reading.

FURTHER READING

Butler, L.J., Gibbons, N., Middleton, C., Elshafie, M.Z.E.B., 2016a. Integrated fibre-optic sensor networks as tools for monitoring strain development in bridges during construction. In: The 19th Congress of IABSE Proceedings, Stockholm, September 21–23, 2016, pp. 1767–1775.

Butler, L.J., Gibbons, N., Ping, H., Elshafie, M.Z.E.B., Middleton, C., 2016b. Evaluating the early-age behaviour of full-scale prestressed concrete beams using distributed and discrete fibre optic sensors. J. Constr. Build. Mater. 126, 894–912.

Davila Delgado, J.M., Butler, L.J., Gibbons, N.J., Brilakis, I., Elshafie, M.Z.E.B., Middleton, C., 2016. Management of structural monitoring data of bridges using BIM. ICE Bridge Engineering themed issue on information technology in Bridge Engineering and Construction. DOI: 10.1680/jbren.16.00013. Available online ahead of print.

CHAPTER FOURTEEN

Repair, Strengthening, and Replacement

14.1 INTRODUCTION

The most common causes of bridge failure are structural and design deficiencies, corrosion, construction and supervision mistakes, accidental overload and impact, scour, and lack of maintenance or inspection (Biezma and Schanack, 2007). To overcome the adverse effects caused by these deteriorated structures, repair and rehabilitation work needs to be carried out from time to time during a bridge's service life.

In recent years, the rapid deterioration of steel bridge structures has become a serious technical and economic problem in many countries, including both developed and developing countries. As shown in Fig. 14.1(A), the bridge built in the United States between the 1920s and 1940s exhibited severe deterioration by the 1980s and resulted in the publications such as "America in Ruins: The Decaying Infrastructure" (Choate and Walter, 1983). In Japan, bridges built during the repaid economic growth period between the 1950s and 1980s began to exhibit deterioration in the 2010s, as can be found in Fig. 14.1(B). Since then, "Japan in ruins" has also become a concern. Taking the bridges older than 50 years (the bridge design service life) in Japan as an example, the ratio has increased remarkably from 6% in 2006 to 20% in 2016, and it is predicted to be around 47% in 2026, as shown in Fig. 14.2.

With aging, deterioration of bridges becomes a serious problem and seriously affects the serviceability and durability of bridges. Therefore, appropriate repair, strengthening, or replacement work should be performed on aged bridge structures to ensure their good performance in service condition. After tens of years' service, these old bridges need to be strengthened integrally for the whole bridges or repaired locally for certain steel members, or replaced. Therefore, bridge inspection, maintenance, rehabilitation, retrofitting, resilience, sustainability has also become a very essential factor in contemporary bridge engineering.

Bridge Engineering
http://dx.doi.org/10.1016/B978-0-12-804432-2.00014-1
245

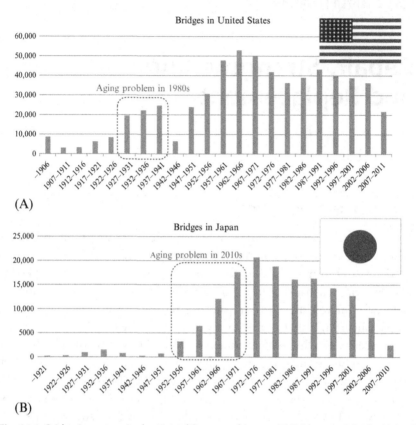

Fig. 14.1 Bridge inventory in the United States and Japan. (A) Bridge stock in the United States. (B) Bridge stock in Japan. *(Courtesy of MLIT.)*

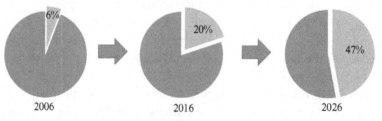

Fig. 14.2 Bridges older than 50 years in Japan. *(Courtesy of MLIT.)*

The repair, strengthening, and replacement are close topics and alternative options for bridge engineers. The decision should be made according to the current condition, predicted deterioration and consequence, and the cost of remedial measures at different stages. The purpose of maintenance and repair activities is to keep bridge structures in functional and safe conditions as the situation permits. In general, the deterioration or destruction can proceed very rapidly once started, and postponed maintenance can result in the development of major repair jobs. Thus, prompt and adequate maintenance is important for aged bridge structures. Considering the relatively high cost for replacing as well as the great impact on the public transportation, repair and strengthening on the aged bridges is generally more preferable (both environmentally and economically) than to demolish and replace them by building new bridges. Nevertheless, bridge replacement is an option in case of severe damages of whole structures and higher cost of repair or strengthening work than replacement or reconstruction. It is generally a hard decision to choose among repair, strengthening, and replacement, but the option could be simplified as: repair now, repair later, strengthen later, or ultimately replace (Ryall et al., 2000).

In this chapter, the commonly used repair and strengthening techniques for steel bridges and concrete bridge will be described.

14.2 REPAIR AND STRENGTHENING OF CONCRETE BRIDGES

14.2.1 Repair of Concrete Structures

The common damages in concrete bridges include the cracks, stripping and exposure of reinforcement, water leakage, flake off, and discoloration, and deterioration, and the corrosion of the reinforcement is one of the main causes. The following matters are of particular interest in the repairing of concrete bridges, including reinforcement cover, carbonation depth, chloride contamination profile, concrete mix details, age of concrete, and environmental factors influencing contamination and condition (Institution of Civil Engineers, 2008).

The typical repair work for concrete bridges includes: crack repair work (e.g., crack injection and filling work), section repair work (e.g., plastering and shotcrete methods), surface treatment (e.g., surface coating), member replacement, peeling prevention, grout reinjection, water proofing work, prestressing installation, and desalination treatment. (Committee on Maintenance and Management of Public Civil Engineering Facilities, 2012).

14.2.2 Strengthening of Concrete Structures

Strengthening work on concrete structures has been a difficult operation. In these circumstances the choice may be limited to the following sections.

14.2.2.1 Traditional Methods

There are several methods to strengthen the concrete members, including (1) increase of the concrete member sections, by widening or heightening; (2) drilling and grouting additional reinforcement, perhaps in the form of stressed bars (Ryall et al., 2000); (3) increase the number of main girders; and (4) increase the number of support points or bearings. However, the development of new techniques and materials in recent years has created new possibilities, such as the plate bonding method and the prestressing method.

14.2.2.2 External Prestressing Method

External prestressing means prestress introduced by tendons locates at outside of a structural member, only connected to the member through deviators and end-anchorages. The main advantages using this technique are higher utilization of small sectional areas, ease in inspection of the tendons and in their replacement and low friction losses. Unlike the other strengthening techniques, external prestressing method is an active strengthening technique. For this reason, it leads to a reduction of deflections, stresses, and cracks in structural members. Therefore, the application of external prestressing is particularly appropriate when there is a need to substantially increase the flexural strength or correct deficient service behavior, e.g., excessive deflection or cracking (Preto, 2014). The external prestressing used for both new bridge construction and existing structures that need to be strengthened due to several reasons such as: changes in use, deficiencies in design or construction phase, and structural degradation (Nordin, 2005). The external prestressing can be applied with either steel strands or FRP tendons, like those used in loading tests as shown in Fig. 14.3. An application example of external prestressing using fiber-reinforced polymer (FRP) tendons in highway bridge construction is shown in Fig. 6.7.

14.2.2.3 Steel Plate (or FRP Sheet) Bonding

Strengthening of concrete structures with bonded steel plate or FRP laminates or sheets is also frequently used. In this method, the plate made of steel or fiber-reinforced composite materials is attached to the concrete surface by bonding and to form a new composite section. The plate provides the concrete section with additional tensile capacity at maximum eccentricity, where it is most effective for resisting the bending moment. This method can also be used for increasing the shear force carrying capacity. As the plate is used as an external member, the subsequent inspection and monitoring

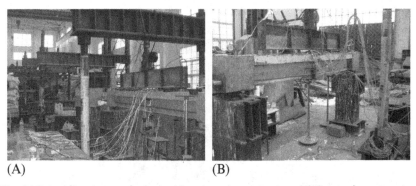

(A) (B)

Fig. 14.3 Loading test on beams with external prestressing. (A) External prestressing (steel strand). (B) External prestressing (FRP tendon). *(Photos by Ding.)*

work are relatively easy. Also, there is very limited section increase by using this method. However, as the effectiveness of plate bonding is entirely dependent on the bonding between the external plate and concrete members, the surface cleaning work and the bonding work must be performed with care. For the case of FRP plate, there are stressed plates bonding and unstressed plate bonding, respectively.

14.3 REPAIR AND STRENGTHENING OF STEEL BRIDGES
14.3.1 Damages in Steel Bridges

The steel bridges are generally very durable if properly maintained. The primary causes of metal bridge deterioration include the corrosion, fatigue, large deformations, and relaxation and drop off of high strength bolt.

14.3.1.1 Corrosion
Corrosion is a common problem for steel bridges, and it is the gradual destruction of steel by chemical reaction with their environment. Corrosion will lead to the loss of the effective materials and reduction of the steel thickness, thus results in the decrease of the stiffness and increase of the structural deformation. In bridge structures, the corrosion is likely to happen at the girder end where water may come from the expansion joint, the top of the bottom flange where soil and dust are easy to accumulate, the bearing surroundings, and connections or joints. Ultrasonic thickness gauges or laser displacement sensors (Fig. 14.4) can be used for the metal thickness measurements.

Fig. 14.4 Metal thickness measurement device with laser displacement sensors. *(Photos by LIN.)*

14.3.1.2 Fatigue

Fatigue failure of steel bridges is another significant problem affecting the remaining service life of existing steel bridges. In general, fatigue can be defined as the weakening of steel materials or accumulation of damage at a localized region caused by cyclic loading or repeatedly applied loads. On this occasion, the material may damage when the nominal maximum stress is still much less than the material strength determined from the material tests. When the material is subjected to the repeated loading above a certain threshold, microscopic cracks will begin to occur at locations in stress concentration. Then the crack will propagate suddenly causing the fracture of the steel members.

14.3.1.3 Large Deformation

The large deformations in steel bridges may happen due to the out-of-plane deformation or buckling of steel members caused by the local stress concentration, overlarge external load, impact, or seismic load. The web without sufficient stiffeners and slender secondary members is vulnerable to such damages.

14.3.1.4 Relaxation and Drop Off of High Strength Bolt

High strength bolts located at the girder end and interconnecting section of bottom flange are easy to suffer such damages due to the water-induced corrosion. Relaxation of high strength bolt may occur on members subjected to vibration loading, or due to the inadequate tightening force during construction.

14.3.2 Repair of Steel Structures

A larger number of parameters should be considered in selecting the most appropriate repair technique for a bridge, such as construction materials, member connection methods like welded, bolted or riveted, degree of redundancy, reasons of deterioration, and potential cost (Institution of Civil Engineers, 2008).

14.3.2.1 Repair of Cracking

The cracking of steel bridge can be caused by many reasons, but fatigue is probably the most common reason in modern steel bridge. This problem is more severe in welded steel bridges. According to the "Manual for Repair and Retrofit of Fatigue Cracks in Steel Bridges" (Dexter and Ocel, 2013), the repair and retrofit techniques for fatigue cracks be divided into three major categories, including (1) surface treatments, (2) repair of through-thickness cracks, and (3) modification of the connection or the global structure to reduce the cause of cracking.

The surface treatment techniques, such as the grinding and impact treatments, are used to improve the welded section and increase the fatigue strength of uncracked welds. Among them, the hammer peening is the commonly used impact treatment method with least cost.

For repairing the through-thickness cracks, there are drilling hole method, adding doubler plates method, and posttension methods. Small fatigue cracks caused by stress concentration can be effectively repaired by drilling a circular hole (also named as stop hole) at the tip of the crack, which is useful for removing the sharp notch at the crack tip and reducing the stress concentration, thus no further propagation will occur. In addition, the "bolting-stop-hole" method (Uchida, 2007) by tightening the stop hope with high strength bolts and the "bolting-stop-hole with attached plates"

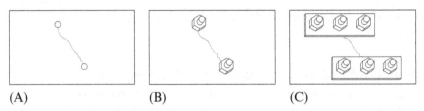

Fig. 14.5 Repair of cracks in steel bridges. (A) Stop-hole. (B) Bolting-stop-hole. (C) Bolting-stop-hole with attached plates.

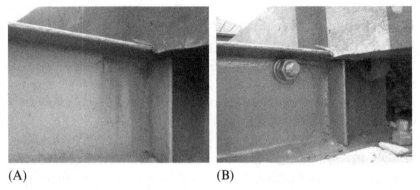

Fig. 14.6 Repair of cracks by bolting-stop-hole method. (A) Before strengthening. (B) After strengthening.

method (Mori et al., 2011) were also proposed, as shown in Figs. 14.5 and 14.6.

14.3.2.2 Adding Doubler Plates

Adding doubler plates method is to add cross-sectional area by using doubler plates, which in turn reduces stress ranges. Doublers can be used to repair cracks due to fatigue or a section that has been heavily damaged by corrosion (Connor et al., 2005). For fatigue cracks on plane members like the flange or the web, the stop hole method can be used as an emergency treatment, and then the doubler plates should be used as an long-term solution. Though doubler plates can be either bolted or welded connection with the existing structural members, doubler plates with high strength bolts are always recommended by considering the fatigue life of steel bridges. In this method, the coating on the steel surface should be removed to ensure the friction on the contacting interface. The image of this method and an application example is shown in Figs. 14.7 and 14.8, respectively.

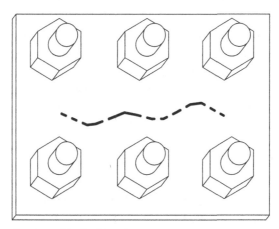

Fig. 14.7 Strengthening with doubler plates.

Fig. 14.8 Doubler plates strengthening examples.

14.3.2.3 Prestressing Method

Similar to the external prestressing method used for strengthening the concrete bridges, this method can also be used for repair or strengthening of steel bridges. The use of prestressing is an active way to change the load or stress distribution and to reduce the displacement. For repairing cracked steel sections due to fatigue, it is recommend that the tip of the existing crack be removed by drilling in addition to applying a posttensioning force. Also, proper corrosion protection must be applied to the posttensioning system (Dexter and Ocel, 2013). The effectiveness of this method in repairing a steel-composite girder with fatigue cracks was also confirmed experimentally (Albrecht and Lenwari, 2008).

14.3.2.4 Member Replacement

In case of the severe damage such as member fracture or large area corrosion, member replacement provides another effective and economical option. However, the stability of the existing structures and the load retribution during member removal and replacement should be carefully examined to ensure the bridge safety.

14.3.3 Strengthening of Steel Structures

In general, most of the repair methods are also used in strengthening of steel bridges; therefore, the methods will not be discussed in detail again in this section. For example, several practical application of prestress in strengthening steel or composite bridges was also reported (Pantura Work Group, 2012), such as the Friarton Bridge (a steel box-girder with lightweight concrete deck) and the Avonmouth Bridge (a twin steel box girder with composite concrete deck). In the Friarton Bridge, external prestress was used to reduce tensile stresses in top flange casused by the increase of external loading. While for the Avonmouth Bridge, which was designed to carry dual three-lane road plus but later strengthened to carry dual four-lane carriageway within the existing overall width. A prestressing tie was installed leading to a saddle near the top of the pier diaphragm in order to reduce the global moments and shear forces within the boxes.

The application of FRP composites for strengthening structural elements has become an efficient option to meet the increased cyclic loads or repair due to corrosion or fatigue cracking (Kamruzzaman et al., 2014). FRP is a relatively new class of composite material manufactured from fibers and resins, and has proven efficient and economical for the development and repair of new and deteriorating structures in civil engineering practice. FRP composite materials possess superior mechanical properties including impact resistance, high strength, low density, high stiffness and flexibility, anticorrosion properties, high durability, and convenient and fast construction, which make them ideal for widespread applications in construction worldwide (Lin et al., 2014a,b). FRP plates, FRP sheets/strips are also effective in the strengthening of steel members to extend their fatigue lifetime. In this method, however, the galvanic corrosion shall be prevented appropriately, and sufficient bonding shall be provided to ensure the long-term behavior of the strengthened steel bridges.

In addition to the strengthening method discussed earlier, there are many other methods which can be used according to the actual conditions of the bridge structures. For instance, for steel girder bridge with a noncomposite reinforced concrete deck, the shear studs may be postinstalled between the steel girders and the concrete deck to increase the load carrying capacity, such as the KY 32 Bridge over Lytles Creek, located in Scott County, Kentucky (Peiris and Harik, 2013). A continuous steel plate girder bridge built in 1939 was strengthened by using additional main girders, which were connected by prestressed tension rods for unloading the existing structure (Harald, 2008). New strengthening methods can be proposed by changing existing structural systems, application of new materials, or adding new structural members.

14.4 BRIDGE REPLACEMENT

It is worth pointing out that not all bridges can be rehabilitated economically, and replacement (or reconstruction) is also an alternative option and could be more cost-effective for seriously damaged bridges. The bridge replacement means a deteriorated superstructure is replaced and upgraded by a new superstructure design with new requirement for load carrying capacity, serviceability, fatigue, etc. (Chen and Duan, 2014). An example is the former Choshi Bridge located at Choshi-shi, Chiba Prefecture, which is close to the river mouth of Tone River connecting Choshi-shi (Chiba Prefecture) and Kamisu-shi (Ibaraki Prefecture) in Japan. It was a five-span truss highway bridge, which extended 406 m in length and 8.2 m in width. The longest span was 107 m and the deck was 7 m in width. The former Choshi Bridge was constructed and opened to traffic in 1962. As located at the saline environment, severe corrosion at the deck system, top and bottom chords as well as the gusset plates at the joints. Though member repair and replacement were conducted several times during the service stage, this bridge was finally demolished in 2011 after the open to traffic of the new cable stayed bridge, as shown in Fig. 14.9.

Bridge replacement or reconstruction is not only due to economic reasons, but sometimes also for inheritance of construction technologies, such as the Kintaikyo Bridge (Fig. 14.10) located at Iwakuni city in Japan. Kintaikyo Bridge is primarily made of wood, and it is vulnerable to natural disaster. As a way of long sustaining the Kintaikyo Bridge, a unique system was established by the City of Iwakuni: instead of reinforcing the existing

Fig. 14.9 The former Choshi Bridge (before removal).

Bridge structure, the City decided to guarantee the succession of Bridge building technology, so as to ensure repeated rebuilding of the Bridge. This solution is indeed unprecedented and unique in the history of bridges in the world (Yoda et al., 2010).

14.5 CASE STUDY: A STRENGTHENING METHOD FOR RAILWAYS BRIDGES IN JAPAN

In Japan, rail transport is a major means of passenger transport between major cities and metropolitan areas, and steel bridges have been widely used in the Japan railway system. However, as many of those bridges have been used in service for tens or even more than one hundred years, many of them need to be strengthened integrally for the whole bridges or repaired locally for certain steel members. Similar to Japan, a survey conducted about the railway bridges in Europe (covering over 220,000 bridges owned by 17 different railways) indicated that more than 35% of the bridges are more than 100 years old (Olofsson et al., 2005, 2007). With the purpose of reducing the stress levels and extend fatigue service life of aged steel bridges, a rehabilitation method for short-span railway, connection between longitudinal and lateral girder in plate girder bridge, and steel columns was proposed. This research was performed by Maebashi Institute of Technology, Waseda University, Taiheiyo Materials Corporation, and Railway Technical Research Institute.

(A)   (B)

(C)

Fig. 14.10 Kintaikyo Bridge reconstruction project. (A) New arch ribs. (B) Falseworks for arch construction. (C) After replacement. *(Photos by Yoda.)*

14.5.1 Strengthening Method Description

The proposed method aims to be used for strengthening short-span steel railway bridge superstructures and longitudinal-lateral beam connection in plate girder bridges. The philosophy of the proposed method is to change the steel section to be the composite section by integrating with new materials, which in turn reduces stress ranges and extend the residual fatigue life of the aged structures. New materials including rubber-latex, rapid hardening concrete, reinforcement, and glass fiber-reinforced polymer (GFRP) plates are used. In the first step, the old structural steel is cleaned and then rubber-latex mortar is sprayed on the surface of the structural steel for protecting the structural steel from corrosion, increasing the bond strength on the steel-concrete interface, and reducing the noise in the service stage. Concrete and mortar, including styrene-butadiene rubber- latex, show various abilities especially in adhesion bonding, waterproofing, and shock absorption and abrasion resistance. In the second step, reinforcing bars and GFRP plates should be installed. GFRP plates are very easy to carry due to their light-weight. In this method, the GFRP plates are used as formworks for concrete casting. Then after that, the light-weight rapid hardening concrete is poured to finish the strengthening. For the maintenance or repairing of the railway bridge, rapid construction and light weight are the critical points. In this strengthening method, the concrete casting can be scheduled in the night time and will not affect the public traffic. The real effects of the present methods were confirmed by following application examples.

14.5.2 Application in Strengthening Short-Span Aged Railway Superstructures

The image of this strengthening method in strengthening short-span aged railway superstructures is shown in Fig. 14.11. By using the present strengthening method, the total procedures generally can be finished within 2 or 3 days without affecting the railway transport. In order to confirm the real effects of this method, both laboratory test and field test were performed.

The old steel railway bridge used for laboratory tests in this study had been used in service for almost 100 years, which was originally built in 1912 with two longitudinal steel girders. The specimen was 4380 mm in length and was simply supported at a span of 3870 mm. After strengthening, the concrete thickness was 200 mm with a width of 1530 mm. GFRP plates with the thickness of 5 mm were used as formworks and shear connection devices with the concrete. Reinforcing bars of D13 nominal diameter were used for both longitudinal and transverse reinforcing bars in the concrete slab. Size dimensions of the test specimen are shown in Fig. 14.12. The

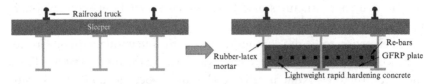

Fig. 14.11 Strengthening of aged short-span steel railway bridges.

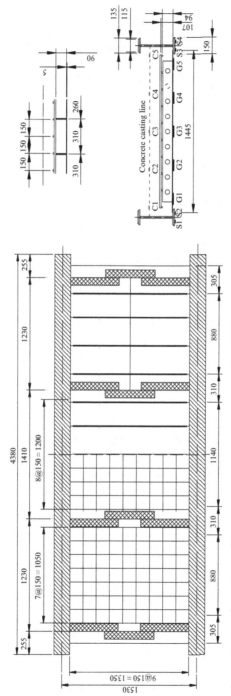

Fig. 14.12 Size dimensions of test specimen for laboratory test (RBL-1& RBL-2).

Fig. 14.13 Strengthening of the laboratory test specimen. (A) Before strengthening. (B) Rubber-latex mortar coating. (C) GFRP and reinforcement arrangement. (D) Concrete casting.

strengthening sequence of the aged steel railway bridge is shown in Fig. 14.13.

The test specimen was supported by a roller system at two ends, with a loading beam in the span center. In order to check the reliability of the measuring equipment and the stability of the test specimen, two levels preloading of 50 kN, 100 kN were applied before the experiment was carried out. The static loading test set-up is shown in Fig. 14.14.

The load-displacement curve obtained from the numerical analyses was compared with the experimental data as shown in Fig. 14.15. The displacement was taken from the vertical deflection at the bottom mid-point of the composite bridge, which was the span center point on bottom surface of the longitudinal beam. Before strengthening, a simple test was performed to generate the load-displacement curve of the original old steel bridge (applied load was only up to 33.8 kN to avoid damage or plasticity of the old steel). For the old steel bridge that has been used around 100 years, fatigue is a major concern. Considering that the design loads are fixed values for railway bridges, so how to increase the bridge stiffness in service stage is the heart of

Fig. 14.14 Laboratory test set-up.

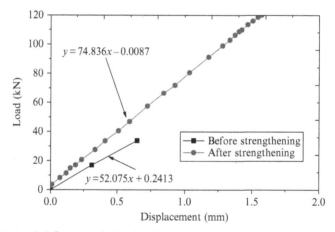

Fig. 14.15 Load-deflection relationship ($P \leq 50$ kN).

this matter. And approximately 43.7% rigidity increase was also confirmed by comparing the slopes of the load versus displacement curves from the test results, indicating that the deformation of the bridge and the stress levels of the steel members can be greatly reduced. Therefore, service life of the aged steel bridge can be extended.

The noise problem is another concern in strengthening aged steel railway bridges. In Japan, there are a large number of such old steel railway bridges, which cannot be replaced by intrinsically more quiet concrete structures due to the increase in weight, cost, and construction height. Therefore, noise reduction was a strategic event in strengthening the old steel railway bridges. In this strengthening method, rubber-latex and concrete members were integrated with the old steel girders not only with the purpose of increasing its stiffness and reducing the stress levels but also with the aim of reducing the noise levels. In order to evaluate the noise reduction effect of the old steel railway bridge after strengthening, the hammer test was performed for the old steel girder and the hybrid girder after maintenance. Vibration accelerations were recorded by using accelerometers arranged on the web in the span center, as shown in Fig. 14.16.

Noise measurement set-up for the old steel railway bridge before and after integration was shown in Fig. 14.16. As for steel girders the structure-borne noise was remarkable when the frequencies from 125 to 2000 Hz, sound pressure levels within this range were recorded. Besides, the all-pass value (AP) was also calculated within this range. The one-third octave filters results shown in Fig. 14.17 indicates that the sound pressure level of 5–15 dB was reduced after strengthening, and about 15 dB sound level reduction can also be confirmed for AP values.

To confirm the real effects of the present strengthening method, an aged steel bridge being used in service was strengthened and tested. The steel railway bridge had been used in service for around 61 years, which was originally built in 1952 with four longitudinal girders. The bridge was 3850 mm in length and was simply supported at a span of 3160 mm. Concrete with the thickness of 200 mm, GFRP plates with the thickness of 5 mm, and reinforcing bars of D13 nominal diameter were employed for strengthening

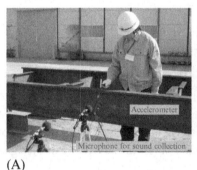 

(A)   (B)

Fig. 14.16 Impact test before and after strengthening. (A) Before integration. (B) After integration.

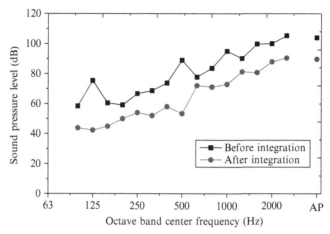

Fig. 14.17 Impact noise before and after strengthening (about 15 dB decrease).

the old railway bridge. In this bridge, only two connecting beams were used at the ends between the longitudinal beams. The field test specimens were denoted by FT-1 (before strengthening) and FT-2 (strengthening), respectively. The size dimensions of the specimen are shown in Fig. 14.18. The aged steel railway bridge before and after strengthening and the strain gauges used in the field test are shown in Fig. 14.19.

In the field tests, six strain gauges in total were used on the steel girder in the mid-span section to measure the flexural strain. Four of them were attached to the lower surface of the bottom flange in mid-span section (ch.1–ch.4), while the other two were used on lower surface of the top flange in the mid-span of two outer girders (ch.5–ch.6), as shown in Fig. 14.18. Train 10,000 series and 11,000 series in Japan Railways were employed as the running trainloads in the field test. The deadweight of train 11,000 series is slightly heavier than that of the 10,000 series train, but the with similar passenger capacity. In the field tests, the flexural strain under five running trains (three of them belong to train 10,000 series and the other two belong to train 11,000 series) was measured on the aged bridge before strengthening. After strengthening, six running trains including three trains of 10,000 series and three trains of 11,000 series were employed in the tests.

The average maximum stress (the measured strain multiplied by the Young's modulus of structural steel) of each main girder under different running trains was summarized in Table 14.1. It is found that in all cases the maximum stress of the strengthened bridges was always much smaller than that of the unstrengthened bridges. For the bridge under the train 10,000,

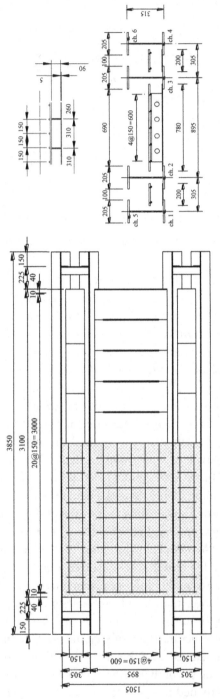

Fig. 14.18 Size dimensions of test specimen for field test (RBF-1).

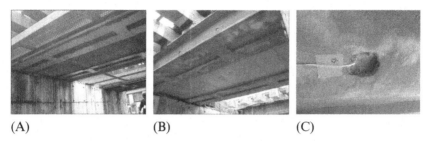

(A)   (B)   (C)

Fig. 14.19 Field test specimen before and after strengthening. (A) Before strengthening. (B) After strengthening. (C) Strain gauges.

Table 14.1 Stress results on Flanges of the Aged Steel Railway Bridges

Specimen	Type	Average Maximum Stress (N/mm^2)					
		ch.1	ch.2	ch.3	ch.4	ch.5	ch.6
FT–1	10,000	15.2	11.5	14.9	11.2	−15.4	−9.3
	11,000	17.3	12.8	17.2	12.7	−17.3	−10.8
FT–2	10,000	12.0	10.5	8.6	8.8	−11.5	−8.1
	11,000	14.4	12.6	10.4	10.4	−14.2	−10.3
Stress reduction (%)	10,000	21.1	8.7	42.3	21.4	25.3	12.9
	11,000	16.8	1.6	39.5	18.1	17.9	4.6

the tensile stress reductions of 21.1%, 8.7%, 42.3%, and 21.4% were confirmed in the four main girders from ch.1 to ch.4, respectively. For the whole section, the maximum tensile stress was reduced by 23.3%. According to MLIT codes, the residual service life of the strengthened bridges can be extended to 2.2 times of the original railway bridge (MLIT, 2009). The compressive stress reduction of 25.3% and 12.9% was also confirmed for the two top flanges. When subjected to train 11,000 series, the average maximum stress reductions of 16.8%, 1.6%, 39.5%, 18.1%, 17.9%, and 4.6% were also confirmed in chapters 1–6, as shown in Table 14.1. The reduction of stress ranges under live load can result in the greatly extension of the residual fatigue service life of the aged steel railway bridges, and the effects of the present strengthening method for short–span railway bridge superstructures were confirmed.

14.5.3 Application in Strengthening Short-Span Aged Railway Superstructures

In steel plate girder bridges, the damage (like crack or fracture) frequently occurs on the longitudinal and lateral beam connections due to fatigue or corrosion, as shown in Fig. 14.20. For steel railway bridges subjected to large train impact and vibration, fatigue is more important than corrosion. Particularly in welded connections, fatigue damages have been frequently reported. On this background, the strengthening method proposed in this study was used to improve the fatigue performance of the connections between longitudinal and lateral beams in steel plate girder bridges. Rubber-latex mortar, GFRP plates, rapid hardening concrete, and reinforcements are used to enhance the stiffness, load carrying capacity, as well as the durability of connections in old steel railway bridges. To confirm the effects of the present strengthening method in strengthening such connections, a specimen was designed according to the real connection in an aged railway bridge in Japan. Each of the specimens was 2.1 m in length and was simply supported at a span length of 2 m. Vertical stiffeners were welded at support sections to prevent buckling failure before flexural failure. The typical geometry and design details of test specimen before and after strengthening are shown in Fig. 14.21.

Depending on the railroad car axle locations, the longitudinal-lateral beam connection can be under either positive bending or negative bending moments. For this reason, two steel connections were used for this study, one connection was subjected to positive bending moment (denotes as SC-P-1 for original steel connection and SC-P-1 for the strengthened connection), and the other one was designed for negative bending moment (denotes as SC-N-1 for original steel connection and SC-N-1 for the strengthened connection). Before strengthening, a static loading test was performed on the original connection joint, the loading was stopped when the strain reached to 800 μ (about 70% of the yield strain) to avoid damage or plasticity of the structural steel. Thereafter, the steel joint was strengthened and loading tests were performed again to confirm the real effects of the present strengthening method. The set-up of the loading tests is shown in Fig. 14.22.

With the purpose of measuring the strain distribution on the welded steel connections, four 1-axis strain gauges and two 3-axis strain gauges were used on the web of the steel connection. Since diagonal fatigue cracks are mainly observed in web corners of the longitudinal beam, the principal (maximum or minimum) strains at those places (PS-1 and PS-2 in Fig. 14.21) are of primary interest in the strain measurement. Under the bending moment, the PS-1 is in tension while the PS-2 is in compression, thus the maximum principal strain at PS-1 and the minimum principal strain at PS-2 are discussed later.

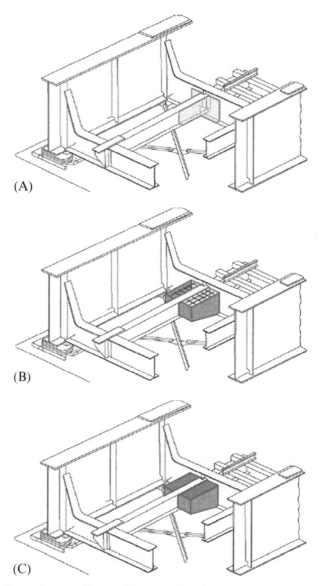

Fig. 14.20 Strengthening of longitudinal-transverse beam connection in aged steel railway bridges (Lin et al., 2014c). (A) Rubber-latex mortar coating. (B) Set-up of GFRP plates & reinforcements. (C) Concrete casting.

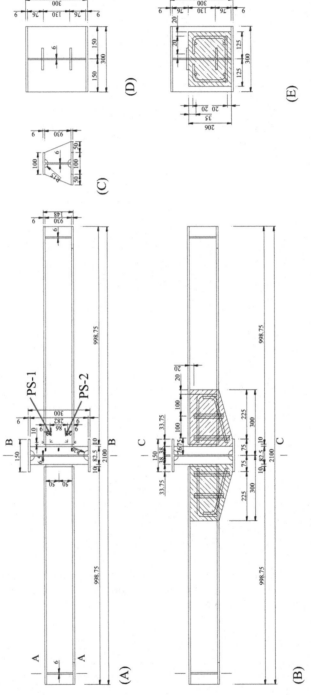

Fig. 14.21 Size dimensions of the steel joint before and after strengthening. (A) Side elevation of the connection before strengthening. (B) Side elevation of the connection after strengthening. (C) A-A. (D) B-B section. (E) C-C section.

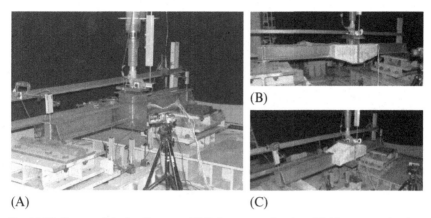

Fig. 14.22 Set-up of the loading test. (A) Before strengthening. (B) After strengthening-positive bending. (C) After strengthening-negative bending.

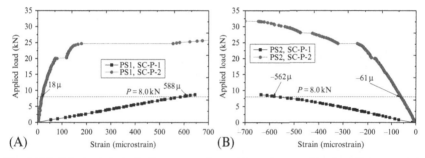

Fig. 14.23 Principal strain in the connection subjected to positive bending moment. (A) Maximum strain at PS-1. (B) Minimum strain at PS-2.

Figs. 14.23 and 14.24 illustrate the principal strain results of the steel connection before and after strengthening. For the connection subjected to positive bending moment (SC-P), taking the applied load of 8 kN as an example, the corresponding maximum principal strain at PS-1 reduced from 588 μ in SC-P-1 to 18 μ in SC-P-2, indicating 97% reduction of the maximum tensile strain. On the other hand, the minimum principal strain at PS-2 reduced from −562 μ in SC-P-1 to −61 μ in SC-P-2, thus 89% reduction can be confirmed.

When subjected to negative bending moment, the principal strain in the original and strengthened steel connection was shown in Fig. 14.24. Taking the applied of 7 kN as an example, the maximum principal strain at PS-1 reduced from 356 μ in SC-N-1 to 55 μ in SC-N-2, and the minimum principal strain at PS-2 reduced from −504 μ in SC-N-1 to −7 μ in SC-N-2.

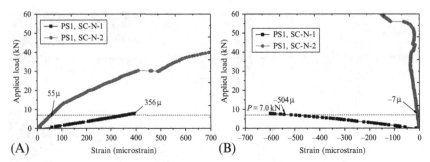

Fig. 14.24 Principal strain in the connection subjected to negative bending moment. (A) PS-1. (B) PS-2.

Therefore, approximately 85% and 99% reduction in the principal strain were confirmed at PS-1 and PS-2, respectively.

Considering the relationship between the stress range (or magnitude) and the stress cycles, the residual fatigue life of the steel connection can be greatly extended, thus the effects of the present method for strengthening such connections can be confirmed. In addition, this method is also applicable for repairing or strengthening of other structures because of the easy obtained construction materials such as GFRP, concrete, reinforcing bars, as well as the easy operational approach.

14.6 EXERCISES

1. Describe the possible "errors" that may cause the failure or collapse of bridge structures.
2. Describe methods for strengthening the concrete structures.
3. Describe the primary causes of metal bridge deterioration and repair methods.

REFERENCES

Albrecht, P., Lenwari, A., 2008. Fatigue strength of repaired prestressed composite beams. J. Bridg. Eng. 13 (4), 409–417.

Biezma, M.V., Schanack, F., 2007. Collapse of steel bridges. J. Perform. Constr. Facil. 21 (5), 398–405.

Chen, W., Duan, L., 2014. Bridge Engineering Handbook, Second Edition: Construction and Maintenance. CRC Press.

Choate, P., Walter, S., 1983. America in Ruins: The Decaying Infrastructure. Duke University Press.

Committee on Maintenance and Management of Public Civil Engineering Facilities, 2012. Bridge maintenance design manual (Draft). Technical report.

Connor, R.J., Dexter, R., Mahmoud, H., 2005. Inspection and Management of Bridges with Fracture-Critical Details Manual for Repair and Retrofit of Fatigue Cracks in Steel Bridges. NCHRP Synthesis 354.

Dexter, R.J., Ocel, J.M., 2013. Manual for repair and retrofit of fatigue cracks in steel bridges. Technical report (No. FHWA-IF-13-020).

Harald, U., 2008. Remarkable Strengthening of an old Steel Highway Bridge, IABSE Congress Report, 17th Congress of IABSE, pp. 200–201.

Institution of Civil Engineers, 2008. ICE Manual of Bridge Engineering: Second Edition. Thomas Telford Ltd.

Kamruzzaman, M., Jumaat, M.Z., Sulong, N.H.R., Islam, A.B.M.S., 2014. A review on strengthening steel beams using FRP under fatigue. Sci. World J. 2014, 702537.

Lin, W., Yoda, T., Taniguchi, N., 2014a. Rehabilitation and restoration of old steel railway bridges: laboratory experiment and field test. J. Bridge Eng. ASCE 19 (5), 04014004.

Lin, W., Yoda, T., Taniguchi, N., Satake, S., Kasano, H., 2014b. Preventive maintenance on welded connection joints in aged steel railway bridges. J. Constr. Steel Res. 92, 46–54.

Lin, W., Yoda, T., Taniguchi, N., Sugino, Y., 2014c. Rehabilitation performance of welded joints in aged railway bridges. J. Struct. Eng. JSCE 60A, 887–896.

Ministry of Land, Infrastructure, Transport and Tourism (MLIT), 2009. Standard specification for railway structures.

Mori, T., Shirai, A., Sasaki, K., Nakanuma, M., 2011. Repair for fatigue cracks at connection between main girder web and lateral girder flange by bolting stop-hole method with attached plates. J. JSCE 67 (3), 493–502.

Nordin, H., 2005. Strengthening Structures With Externally Prestressed Tendons: Technical report. Luleå University of Technology.

Olofsson, I., Elfgren, L., Bell, B., Paulsson, B., Niederleithinger, E., Jensen, J.S., Feltrin, G., Taljsten, B., Cremona, C., Kiviluoma, R., Bien, J., 2005. Assessment of European railway bridges for future traffic demands and longer lives – EC project: sustainable bridges. Struct. Infrastruct. Eng. 1 (2), 93–100.

Olofsson, et al., 2007. Sustainable bridges—a European integrated research project. Background and overview. In: Proceeding of Sustainable Bridges—Assessment for Future Traffic Demands and Longer Lives, pp. 29–49.

Pantura Work Group, 2012. Strengthening and Repair of Steel Bridges-Case Studies.

Peiris, A., Harik, I., 2013. Field testing of steel bridge girders strengthened using ultra high modulus (UHM) carbon fiber reinforced polymer (CFRP) laminates. In: Fourth Asia-Pacific Conference on FRP in Structures (APFIS 2013), December 11–13, 2013. Melbourne, Australia.

Preto, P.B., 2014. Guidelines for External Prestressing as Strengthening Technique for Concrete Structures. Departamento de engenharia civil of Instituto Superior Técnico.

Ryall, M.J., Parke, G.A.R., Harding, J.E., 2000. The Manual of Bridge Engineering. Thomas Telford Press.

Uchida, D., 2007. Fatigue strength evaluation of out-of-plane gusset welded joints repaired by bolting-stop-hole method. Mitsui Engineering & Shipbuilding Technology, No. 190 (2007), 45–53.

Yoda, T., Kasano, H., Lin, W., 2010. Sustainable technology of the Japanese Historical Timber Bridge: Kintaikyo Bridge. In: Proceedings of the 3rd International Conference on Preservation and Research of Historic Bridges, pp. 25–33.

INDEX

Note: Page numbers followed by *f* indicate figures, and *t* indicate tables.